Leishmania Parasites - Epidemiology, Immunopathology and Hosts

Edited by Fernando Almeida-Souza, Flávia de Oliveira Cardoso, Ana Lucia Abreu-Silva and Kátia da Silva Calabrese

Published in London, United Kingdom

Leishmania Parasites - Epidemiology, Immunopathology and Hosts
http://dx.doi.org/10.5772/intechopen.105279
Edited by Fernando Almeida-Souza, Flávia de Oliveira Cardoso, Ana Lucia Abreu-Silva and Kátia da Silva Calabrese

Contributors
Allana Barros Freitas, Ana Lucia Abreu-Silva, Asma Ashraf, Azhar Rafique, Carlos Eduardo Corbett, Claudia Maria Gomes, Deepak Kumar Arya, Fernando Almeida-Souza, Fernando T. Silveira, Flávia de Oliveira Cardoso, Ghulam Rahim Awab, Helen Silva Ribeiro, Intzar Ali, Kaushal Kumar Mahto, Kátia da Silva Calabrese, Luciana V. Lima, Marliane B. Campos, Mohan Kumar, Muhammad Shahid Mahmood, Márcia D. Laurenti, Patrícia K. Ramos, Pooja Prasad, Prabir K. Chatterjee, Rajan R. R Patil, Salma Sultana, Sandra Alves Araújo, Sayydah Surrya Sani, Silvia F. Müller, Stephen Kyari, Tatiane Aranha da Penha-Silva, Tayyaba Sultana, Thiago V. dos Santos, Vania Lucia da Matta, Vikram Vohra

First published in London, United Kingdom, 2024 by IntechOpen
IntechOpen is the global imprint of INTECHOPEN LIMITED, registered in England and Wales, registration number: 11086078, 5 Princes Gate Court, London, SW7 2QJ, United Kingdom

British Library Cataloguing-in-Publication Data
A catalogue record for this book is available from the British Library

Additional hard and PDF copies can be obtained from orders@intechopen.com

Leishmania Parasites - Epidemiology, Immunopathology and Hosts
Edited by Fernando Almeida-Souza, Flávia de Oliveira Cardoso, Ana Lucia Abreu-Silva and Kátia da Silva Calabrese
p. cm.
Print ISBN 978-1-83768-310-9
Online ISBN 978-1-83768-311-6
eBook (PDF) ISBN 978-1-83768-312-3

For EU product safety concerns:
IN TECH d.o.o., Prolaz Marije Krucifikse Kozulić 3, 51000 Rijeka, Croatia,
info@intechopen.com or visit our website at intechopen.com.

Meet the editors

Dr. Fernando Almeida-Souza is a veterinarian with a master's degree in animal science from the State University of Maranhão, Brazil, and a doctorate in cellular and molecular biology from Oswaldo Cruz Institute/FIOCRUZ (IOC/FIOCRUZ), Brazil. He is currently a Professor of Immunology at Post-graduate in Animal Science, at the State University of Maranhão. He also collaborates at IOC/FIOCRUZ in the Laboratory of Protozoology, guiding doctoral and master students. He has several publications in international journals to his credit. He is also a journal reviewer and editor. At present, much of Dr. Almeida-Souza's research is directed towards developing new therapeutic strategies for trypanosomatid diseases, evaluating synthetic and natural product compounds, in addition to studies of cellular ultrastructure, biochemical and cellular and molecular biology assays on tumor cells, pathogenic microorganisms, and experimental biological models.

Dr. Flávia de Oliveira Cardoso is a biologist with a Ph.D. in Parasite Biology from Oswaldo Cruz Institute/FIOCRUZ (IOC/FIOCRUZ), Brazil, where she is a public health technician. She is also a quality manager in the Laboratory of Protozoology, IOC/FIOCRUZ. She researches the immunopathology, diagnosis, and epidemiology of leishmaniasis and Chagas disease as well as drug discovery for their treatment. Dr. Cardoso supervises undergraduate and graduate students. She has several peer-reviewed research articles and book chapters to her credit. She is an editorial board member for *Frontiers in Cellular and Infection Microbiology* and *Biology* as well as a journal reviewer. She is a collaborating professor and researcher at the State University of Maranhão, Brazil, and an invited teacher at Souza Marques University, Brazil.

Dr. Ana Lucia Abreu-Silva started a position as an assistant professor at the State University of Maranhão, Brazil, in 1990, and has had a successful career in research, teaching, and extension in veterinary medicine. She has played fundamental roles in the development of research and graduate studies in Maranhão state. Recently, she began work on an innovative project that aims to track cases of leishmaniasis. She was previously coordinator of the Graduate Program of Animal Sciences at the State University of Maranhão. She is currently a coordinator of the veterinary medicine course at the same institution, as well as a member of the Ethics Committee for Research with Human Beings and a researcher at the Brazilian National Council for Scientific and Technological Innovation.

Dr. Katia da Silva Calabrese is a post-doctoral student at Hôpital Saint-Louis and Université Denis Diderot, Paris. She obtained a master's degree and Ph.D. in Parasitic Biology from the Oswaldo Cruz Institute/FIOCRUZ (IOC/FIOCRUZ), Brazil. She also holds a degree in Biological Sciences from UGF. She is a researcher in public health, the head of the Laboratory of Protozoology, and the coordinator of the Reference Service for Canine Tegumentary Leishmaniasis at IOC/FIOCRUZ. She works in protozoology, especially in Chagas disease, leishmaniasis, and toxoplasmosis, with research projects that include developing formulations for the treatment of protozoa and inflammatory processes; studying resistance and susceptibility mechanisms and interactions of pathogens with their host cells; and immunogenicity, epidemiology, and immunopathology. Dr. Calabrese is a postgraduate advisor at IOC, UEC, UFMG, and UEMA and a member of the IOC Technical Reference Chamber.

Contents

Preface

For many years, leishmaniasis cases were basically restricted to developing countries. As a disease of extreme poverty, measures to control the insect vector, allocating resources for research, and protecting the environment were not adopted by government policies. The rise of unrestricted deforestation, globalization, and co-infections contributed to the spread of the disease to other countries in less than a century. Leishmaniasis is now a category Class I disease according to the World Health Organization (WHO) and is considered an uncontrolled emerging disease. These types of neglected diseases present many challenges, such as a lack of vector control strategies and vaccines for the most pathogenic *Leishmania* strains, parasite resistance, and ineffective treatment for canine leishmaniasis, which makes it increasingly difficult to control these diseases. This book discusses different aspects of *Leishmania* parasites. Chapters are organized into three sections: "Introduction", "Epidemiology and Immunopathology of Leishmaniases" and "New Insights on Leishmaniasis in Domestic Animals".

Section 1 contains the introductory chapter. The six chapters that follow in section 2 "Epidemiology and Immunopathology of Leishmaniases" focus on the epidemiology of leishmaniasis. Chapter 2 reviews the epidemiology of leishmaniasis, Chapter 3 discusses the epidemiology of visceral leishmaniasis in India, and Chapter 4 presents an interesting analysis of leishmaniasis epidemiology and psychosocial aspects of the disease. Chapter 5 is a review of cutaneous leishmaniasis, while Chapter 6 discusses visceral leishmaniasis, including its status, geographical distribution and transmission. Chapter 7 presents an in-depth review of the importance of species-specific leishmania antigens.

Section 3 includes two chapters that discuss leishmaniasis in domestic animals. Chapter 8 focuses on feline leishmaniasis, while Chapter 9 reviews the association between canine leishmaniasis and malignant disorders.

All chapters provide a unique perspective, resulting from the efforts of numerous leading researchers around the world. Overall, the book contributes to a better understanding of *Leishmania* parasites as well as aspects related to epidemiology, immunopathology, and challenges of the disease in humans and domestic animals.

Fernando Almeida-Souza and Ana Lucia Abreu-Silva
State University of Maranhão,
São Luís, Brazil

Flávia de Oliveira Cardoso and Kátia da Silva Calabrese
Oswaldo Cruz Institute,
Rio de Janeiro, Brazil

Section 1

Introduction

Chapter 1

Introductory Chapter: *Leishmania* Parasites – Epidemiology and Immunopathogenesis

Fernando Almeida-Souza, Ana Lucia Abreu-Silva, Kátia da Silva Calabrese and Flávia de Oliveira Cardoso

1. Introduction

Leishmaniases are a group of diseases caused by protozoan parasites from over 20 species belonging to subfamily Leishmaniinae [1, 2]. The parasites are transmitted by over 90 species phlebotomine sandflies from one host to the other by the bite of an infected female [3].

Clinical manifestations of leishmaniasis range from self-healing cutaneous lesions to potentially fatal visceral form. All clinical manifestations depend on the parasite species and the host's genetic and immunological response [4].

These pathologies affect predominantly the poorest people, mainly in tropical countries, exerting a high impact on mortality and morbidity rates. The risk factors for the development of the disease are associated with malnutrition, socioeconomic conditions, population mobility, environmental changes, and climate changes [3].

Due to the high global incidence and prevalence rates, mainly in developing countries, leishmaniases are among the 10 most neglected diseases (ND) in the world [5]. Worldwide, 99 countries are endemic to Leishmaniasis, where more than 12 million people are infected, and 20,000 to 30,000 deaths occur annually [3, 5]. Cutaneous leishmaniasis (CL) is endemic in 89 countries, visceral leishmaniasis (VL) in 80 countries, and both clinical forms are endemic in 71 countries. In the Americas, Brazil is the main country endemic for both CL and VL. Leishmania-HIV coinfection has intensified the burden of leishmaniasis, which have been reported by 42 countries around the world [3, 5].

These neglected diseases present many challenges, such as a lack of vector control strategies, different parasite strains, parasite resistance, and canine leishmaniasis treatment with low effectiveness, which makes it increasingly difficult to control these diseases.

2. Experimental leishmaniasis

Leishmaniasis affects a wide variety of hosts, whether they are natural reservoirs or "accidental" hosts in the *Leishmania* life cycle [3, 6].

IntechOpen

Proposing an experimental model that can demonstrate what happens in both the wild host and the accidental host is a major challenge. The use of rodents as mice (*Mus* spp.) and hamster (*Cricetinae* spp.), as an experimental model, has been proposed as a good alternative to study the *Leishmania*-host relationship. Other models as dogs and primates are also used for leishmaniasis study, principally to test the efficacy of leishmanial vaccines [7–9]. In fact, there are a plethora of animal models for leishmaniasis study each of them with their peculiarities, advantages, and disadvantages. Mice with different genetic backgrounds are used for experimental studies involving parasites of the Leishmania genus, mainly susceptibility and resistance studies [4]. Different response profiles to the parasite infection are observed, and it is possible to classify these mice as sensitive, resistant, and intermediate. The hamster model is commonly used to study cutaneous leishmaniasis caused by *L. brasiliensis*, but it can also be used to isolate species that are difficult to cultivate [8]. The choice of model is related to which individual will provide the best answer to the question the researcher proposed. In leishmaniasis studies, mice and hamster are most commonly used [10]. There is, however, no consensus on the best animal model to be used, but regardless of the model chosen for the study, it is necessary to keep in mind animal welfare, the concept of the 3Rs, and approval of the experimental protocol by the Ethics Committee of Animal Use.

3. Visceral leishmaniasis

Visceral leishmaniasis (VL) is a life-threatening and tropical neglected disease, one of the world's six priority endemic diseases. In the New World, the protozoan Leishmania infantum infects several mammal species, including humans and dogs, which causes the visceral form. In urban areas, canines play a role in the transmission and maintenance of the parasite, which is evidence of a temporal correlation between the occurrence of human and canine visceral leishmaniasis [11]. In several countries, it is an uncontrolled disease and a severe public health issue. To establish public measures for the prophylaxis of this disease, screening the canine population and infected dogs is one of the primary forms of control of VL.

In human visceral leishmaniasis, the liver function and morphology are modified due to inflammatory reactions in the portal and intralobular region that may be locally concentrated or diffusely localized. Besides the involvement of this organ, the parasite could lead to augmenting the spleen; thus, hepatosplenomegaly is a frequent clinical sign. Other clinical signs described are irregular bouts of fever, weight loss, enlargement of the spleen and liver, and anemia [3].

Unlike humans, which develop only visceral form, in canines, the first clinical signs observed are skin lesions such as periorbital alopecia, exudative dermatitis, hyperkeratosis of the footpad, and onychogryphosis. Other frequent signs are hepatosplenomegaly, enlargement of lymph nodes, weight loss, cachexia, epistaxis, pancytopenia, hypergammaglobulinemia, ocular lesions, and chronic renal failure [12–15].

4. Treatment, an urgent need for new drugs

Historically, several treatments against leishmaniasis have been used in the past, such as quinine, arsenicals, and turpentine injection. With variable results, none of them had proven effectiveness. In 1912, the doctor Gaspar Viana used the tartar emetic, trivalent antimonial (SbIII), for the first time successfully in the therapy of ATL [16]. Between 1920 and 1930, pentavalent antimonials (SbV) were introduced in the treatment of visceral leishmaniasis, reducing treatment from 3 to 4 months to weeks. In the 1940s, new formulations of pentavalent antimonials appeared, which are less toxic and still used today: meglumine antimoniate, Glucantime®, and sodium stibogluconate, Pentostam® [17].

Despite the few advances in research on antimonials in recent decades, they are still the first-choice therapy against leishmaniasis. The treatment is long, with intramuscular administration, which sometimes requires hospitalization, causing discomfort to the patient, and high costs [18]. The other drugs available for leishmaniasis treatment are mainly amphotericin B and miltefosine.

Amphotericin B is an antifungal produced by culturing the actinomycetes *Streptomyces nodosus* used to combat systemic infections and is presented as amphotericin deoxycholate and in liposomal form. Liposomal amphotericin B was first used in the treatment of ATL by Sampaio & Mardsen [19]. It is the drug of choice in cases of parasite resistance to antimonials [20].

Miltefosine is an alkylphosphocholine originally developed as an antineoplastic chemotherapy drug. In the 1980s, during its evaluation for this therapeutic purpose, its potent action against leishmaniasis was discovered. In 2002, it was registered in India as the first oral treatment for visceral leishmaniasis [21]. Studies using miltefosine in the treatment of the cutaneous form of the disease resulted in good efficacy against *L. (V.) panamensis*, but for *L. (V.) braziliensis* there was no adequate efficacy [22].

Several active compounds have been tested against leishmaniasis, such as ketoconazole [23], itraconazole [24], fluconazole [25], terbinafine [26], and azithromycin [27], all with variable results. Thermal treatments, such as thermotherapy [28] and cryotherapy [29] have also been tested in the treatment of skin lesions; however, the lack of specificity and the need for professionals trained in the technique makes it difficult to advance the treatment. The search for new drugs should not be discontinued, as the disease progresses, and patients continue to wait for effective and low-toxicity medications.

Acknowledgements

This research was funded by Coordenação de Aperfeiçoamento de Pessoal de Nível Superior – CAPES (Finance Code 001), by Fundação Carlos Chagas Filho de Amparo à Pesquisa do Estado do Rio de Janeiro (26/211.680/2021) and Fundação de Amparo à Pesquisa e Desenvolvimento Científico e Tecnológico do Maranhão (APP-12233/22). HSR (CNPq 150336/2023-3) and FA-S (FAPERJ E-26/203.513/2023) are postdoctoral research fellows. KC (315225/2021-1) and ALA-S (313348/2021-9) are research productivity fellows by Conselho Nacional de Desenvolvimento Científico e Tecnológico - CNPq.

Author details

Fernando Almeida-Souza[1,2*], Ana Lucia Abreu-Silva[1], Kátia da Silva Calabrese[2] and Flávia de Oliveira Cardoso[2]

1 Program of Animal Science, State University of Maranhão, São Luís, Brazil

2 Laboratory of Protozoology, IOC-Fiocruz, Rio de Janeiro, Brazil

*Address all correspondence to: fernandoalsouza@gmail.com

References

[1] Espinosa OA, Serrano MG, Camargo EP, Teixeira MMG, Shaw JJ. An appraisal of the taxonomy and nomenclature of trypanosomatids presently classified as Leishmania and Endotrypanum. Parasitology. 2018;**145**(4):430-442. DOI: 10.1017/S0031182016002092

[2] Kostygov AY, Yurchenko V. Revised classification of the subfamily Leishmaniinae (Trypanosomatidae). Folia Parasitologica. 2017;**64**:020. DOI: 10.14411/fp.2017.020

[3] World Health Organization. Leishmaniasis. 2023 [Accessed: November 27, 2023]. Available from: https://www.who.int/news-room/fact-sheets/detail/leishmaniasis

[4] Cardoso FO, Zaverucha-do-Valle T, Almeida-Souza F, Abreu-Silva AL, Calabrese KDS. Modulation of cytokines and extracellular matrix proteins expression by Leishmania amazonensis in susceptible and resistant mice. Frontiers in Microbiology. 2020;**11**:1986. DOI: 10.3389/fmicb.2020.01986

[5] Pan American Health Organization. Leishmaniasis. 2023 [Accessed: November 27, 2023]. Available from: https://www.paho.org/en/topics/leishmaniasis

[6] Labbé F, Abdeladhim M, Abrudan J, Araki AS, Araujo RN, Arensburger P, et al. Genomic analysis of two phlebotomine sand fly vectors of Leishmania from the New and Old World. PLOS Neglected Tropical Diseases. 2023;**17**(4):e0010862. DOI: 10.1371/journal.pntd.0010862

[7] Gupta S, Nishi. Visceral leishmaniasis: Experimental models for drug Discovery. Indian Journal of Medical Research. 2011;**133**:27-39

[8] Loría-Cervera EN, Andrade-Narváez FJ. Review animal models for the study of leishmaniasis immunology. Revista Instituto Medicina Tropical São Paulo. 2014;**56**(1):1-11. DOI: 10.1590/s0036-46652014000100001

[9] van der Ende J, Schallig HDFH. Leishmania Animal Models Used in Drug Discovery: A Systematic Review. Animals 2023;**13**:1650. DOI: 10.3390/ ani13101650

[10] Nieto A, Domínguez-Bernal G, Orden JÁ, De La Fuente R, Madrid-Elena N, Carrión J. Mechanisms of resistance and susceptibility to experimental visceral leishmaniosis: BALB/c mouse versus syrian hamster model. Veterinary Research. 2011;**42**(1):39. DOI: 10.1186/1297-9716-42-39

[11] Braz MS, Sáfadi T, Ferreira RA, Morais MHF, Silva Z, Rocha CMBMD. Temporal relationship between human and canine visceral leishmaniasis in an urban area in southeastern Brazil: An application of the ARIMAX model. Preventive Veterinary Medicine. 2023;**215**:105921. DOI: 10.1016/j.prevetmed.2023.105921. Epub 2023 Apr 12

[12] Ciaramella P, Oliva G, De Luna R, Ambrosio R, Cortese L, Persechino A, et al. A retrospective clinical study of canine leishmaniasis in 150 dogs naturally infected by Leishmania infantum. Veterinary Record. 1997;**141**:539-543. DOI: 10.1136/vr.141.21.539

[13] Amusategui I, Sainz A, Rodríguez F, Tesouro MA. Distribution and relationships between clinical and

biopathological parameters in canine leishmaniasis. European Journal of Epidemiology. 2003;**18**(2):147-156. DOI: 10.1023/A:1023090929302

[14] Calabrese KS, Cortada VM, Dorval ME, Souza Lima MA, Oshiro ET, Souza CS, et al. Leishmania (Leishmania) infantum/chagasi: Histopathological aspects of the skin in naturally infected dogs in two endemic areas. Experimental Parasitology. 2010;**124**(3):253-257. DOI: 10.1016/j.exppara.2009.10.005. Epub 2009 Oct 23

[15] Diriba WW, Gebremedhin EZ. Assessment of knowledge, attitude, and practices towards canine visceral leishmaniasis based on the one health concept in Weliso and Ejaji towns, Oromia, Ethiopia. Scientific Reports. 2023;**13**(1):20765. DOI: 10.1038/s41598-023-47340-0

[16] Vianna G. Tratamento da leishmaniose tegumentar por injeções intravenosas de tártaro emético. In: Anais do 7° Congresso Brasileiro de Medicina e Cirurgia. Vol. 4. 1912. pp. 426-428

[17] Murray HW. Treatment of visceral leishmaniasis (kala-azar): A decade of progress and future approaches. International Journal of Infectious Diseases. 2000;**4**:158-177

[18] Grimaldi G, Tesh RB. Leishmaniases of the New World: Current concepts and implications for future research. Clinical Microbiology Reviews. 1993;**6**:230-250

[19] Sampaio RN, Marsden PD. Treatment of the mucosal form of leishmaniasis without response to glucantime, with liposomal amphotericin B. Revista da Sociedade Brasileira de Medicina Tropical. 1997;**30**:125-128

[20] Torre-Cisneros J, Prada JL, Villanueva JL, Valverde F, Sánchez-Guijó P. Successful treatment of antimony-resistant cutaneous leishmaniasis with liposomal amphotericin B. Clinical Infectious Diseases. 1994;**18**:1024-1025

[21] Sundar S, Murray HW. Availability of miltefosine for the treatment of kala-azar in India. Bulletin of the World Health Organization. 2005;**83**:394-395

[22] Soto J, Arana BA, Toledo J, et al. Miltefosine for new world cutaneous leishmaniasis. Clinical Infectious Diseases. 2004;**38**:1266-1272

[23] Saenz RE, Paz H, Berman JD. Efficacy of ketoconazole against Leishmania braziliensis panamensis cutaneous leishmaniasis. The American Journal of Medicine. 1990;**89**:147-155

[24] Al-Fouzan AS, Al Saleh QA, Najem NM, Rostom AI. Cutaneous leishmaniasis in Kuwait. Clinical experience with itraconazole. International Journal of Dermatology. 1991;**30**:519-521

[25] Alrajhi AA, Ibrahim EA, De Vol EB, Khairat M, Faris RM, Maguire JH. Fluconazole for the treatment of cutaneous leishmaniasis caused by Leishmania major. The New England Journal of Medicine. 2002;**346**:891-895

[26] Bahamdan KA, Tallab TM, Johargi H, et al. Terbinafine in the treatment of cutaneous leishmaniasis: A pilot study. International Journal of Dermatology. 1997;**36**:59-60

[27] Sampaio RN, Lucas IC, Costa Filho AV. The use of azythromycin and N-methyl glucamine for the treatment of cutaneous Leishmaniasis caused by Leishmania (Leishmania) amazonensis in C57BL6 mice. Anais Brasileiros de Dermatologia. 2009;**84**:125-128

[28] Velasco-Castrejon O, Walton BC, Rivas-Sanchez B, et al. Treatment of cutaneous leishmaniasis with localized current field (radio frequency) in Tabasco, Mexico. The American Journal of Tropical Medicine and Hygiene. 1997;**57**:309-312

[29] Asilian A, Sadeghinia A, Faghihi G, Momeni A. Comparative study of the efficacy of combined cryotherapy and intralesional meglumine antimoniate (Glucantime) vs. cryotherapy and intralesional meglumine antimoniate (Glucantime) alone for the treatment of cutaneous leishmaniasis. International Journal of Dermatology. 2004;**43**:281-283

Section 2

Epidemiology and Immunopathology of Leishmaniases

Chapter 2

Epidemiology of Leishmaniasis

Stephen Kyari

Abstract

A zoonotic illness of importance to the public's health is leishmaniasis. *Leishmania donovani*, *Leishmania tropica*, *Leishmania major*, *Leishmania infantum*, *Leishmania chagasi*, *Leishmania mexican*, and *Leishmania braziliensis* are the most recognised and widely distributed leishmania parasite species, and they are also the ones that cause the disease. On every continent and in more than 90 countries, the disease is present, however it appears to be absent from Australia. The disease is spread by Phlebotomus sandflies, and people, rodents, and other domestic animals act as reservoirs and unintentional hosts. Cutaneous leishmaniasis, diffuse cutaneous leishmaniasis, mucocutaneous leishmaniasis, and visceral leishmaniasis are the four ways the illness can present. The spread of the disease, as well as its appearance and reemergence, are caused by risk factors include regional warfare and wars, political instability, migration of people, substandard housing, climate, vegetation cover, p7oor socioeconomic standard of life, and lack of access to quality medical care. To eradicate the disease, particularly in poor nations where it is still a threat, there is a need for increased public education, government involvement, proper surveillance, and disease reporting.

Keywords: epidemiology, distribution, risk factors, Phlebotomus sandfly, leishmaniasis

1. Introduction

Leishmaniasis is a disease that is caused by the parasite Leishmania; comprising of at least 20 species that affect humans. The parasite is primarily transmitted by the sand flies that belongs to the Phlebotomus in the old world and Lutzomyia in the new world; however, other *principal hosts* are rodents such as *Rhombomys opimus, Tatera indica, Meriones hurrianae, and Meriones libycus gerbils which are of the Zoonotic cutaneous leishmaniasis (ZCL)* [1–5]. *The disease is associated* with and limited by the geographical distribution of its sand fly vectors. Fifteen geographical entities have been classified worldwide, of which 13 zoonotic in nature. In recent years, there have been significant increase in the number of regions becoming *Leishmania*-endemic, accompanied with an increased number of animal and human cases. These parasites have been reported to be maintained and circulated in mammals with each species having a specific reservoir host [6]. The clinical spectrum of the disease in humans can be asymptomatic infections in some individual and can result in high mortality, which depends on the form of the disease and can be categorised into three (4) namely; visceral leishmaniasis (VL), cutaneous leishmaniasis (CL), *diffuse cutaneous leishmaniasis (DCL)* and mucocutaneous leishmaniasis (MCL). In dogs, *Leishmania infantum* and *Lieshmania chagasi are* now regarded as synonyms *and are the species known* to be responsible for chronic visceral-cutaneous disease in this host (canine leishmaniosis). Asymptomatic infections occur in dogs and other canines thereby maintaining the presence of the parasite in endemic regions [6]. The ability of the host macrophages to effectively destroy the parasite is as a result of host- parasite interaction which is responsible for clinical appearance. Leishmaniasis is classified as a prominent neglected disease of the tropics due to the number of death per year (40,000) with a global distribution with Indian sub-continent having the highest number of cases followed by Bangladesh, Sudan, South Sudan, Ethiopia, and Brazil [7, 8].

1.1 Cutaneous leishmaniasis

The aetiological agent responsible for cutaneous leishmaniasis are as follows: *Leishmania tropica*, *Leishmania major*, and *Leishmania aethiopica*, also known as old- and new-world leishmania species, are found in southern Europe, the Middle East, Asia, and Africa, respectively. These species include *Leishmania guyanensis*, *L. viannia braziliensis*, *L. viannia amazonesis*, *L. viannia panamensis*, and the *Leishmania mexicana* complex (Latin America). Skin ulcers on the exposed body parts, such as the face, arms, and legs, as well as lesions that leave the patient permanently disfigured and prone to social judgement, are the hallmarks of cutaneous leishmaniasis. The incubation time is roughly 2 to 6 weeks [9]; cases of up to 3 years have been documented for Old World cutaneous leishmaniasis [10], whereas the incubation period for New World cutaneous leishmaniasis is typically 2 to 8 weeks [11]. Over 85% of newly discovered CL cases in 2018 that were reported to the WHO originated in Afghanistan, Algeria, Bolivia, Brazil, Colombia, Iran, Iraq, Pakistan, Syria, and Tunisia, while over 95% of newly discovered LV cases originated in Brazil, China, Ethiopia, India, Iraq, Kenya, Nepal, Somalia, and Sudan. Finally, four nations—Brazil, Bolivia, Ethiopia, and Peru—reported almost 90% of all new cases of mucocutaneous leishmaniasis (MCL) [12].

1.2 Mucocutaneous leishmaniasis

Leishmania viannia braziliensis, *leishmania viannia amazonesis*, *leishmania viannia panamensis*, and *leishmania viannia guyanesis* are the leishmania species that cause mucocutaneous leishmaniasis. The progression of this illness is controlled by parasite virulence and host cell mediated immunity. The condition starts with erythema and ulceration of the nose, which progresses to continuous degradation of the cartilage in the upper airway and facial structures, oronaso-pharyngeal mucosa, disfigurement, and obstruction of the airway [13].

1.3 Diffuse cutaneous leishmaniasis

Even in nations where leishmaniasis is endemic, diffuse cutaneous leishmaniasis is a rare manifestation of the illness. *Leishmania mexicana* and *Leishmania aethiopica* are the parasites that cause this type of leishmaniasis. A small, painless bump at the site of the innoculation causes spreading, non-ulcerating macules, nodules, and plaques that range in colour from erythematous to violet. The face, upper and lower limbs, and buttocks are the area most commonly affected [14, 15].

1.4 Visceral leishmaniasis

Leishmania donovani is the cause of visceral leishmaniasis. The most serious type of the disease, which affects internal organs like the spleen and liver and can

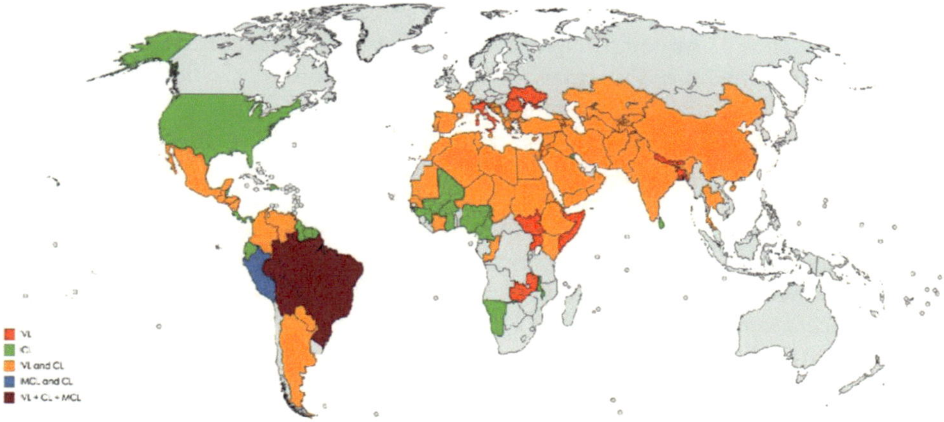

Figure 1.
Shows the geographical distribution pattern of leishmaniasis [20].

	Leishmania species
Old world cutaneous leishmaniasis	*Leishmania tropica, Leishmania major, Leishmania infantum, Leishmania aetheiopica*
New world cutaneous leishmaniasis	*Leishmania vianna* subgenous (e.g., *Leishmania vianna braziliensis, Leishmania vianna panamensis, Leishmania vianna guyanensis, Leishmania vianna peruviana, Leishmania amazonensis*) and *Leishmania Mexicana* complex

Table 1.
Shows old world and new world cutaneous leishmaniasis [21–23].

be deadly if neglected, is Kala-azar cutaneous leishmaniasis, which ranges from hypopigmented macules to infiltrating papules [16]. The incubation period ranges from three to eight months [17]; symptoms include fever, weight loss, hepatosplenomegaly, lymphadenopathy, pancytopenia, and hypergammaglobulinaemia; skin pigmentation may also be present [18]; "kala azar," also known as black disease, may be asymptomatic and self-resolving, but typically has a chronic course and can be fatal with or without treatment [19]; death typically results due to bacterial infections. In the Indian subcontinent, Bangladesh, India, and Nepal are the most severely affected and account for 28% of cases, while Brazil contributes 20%. The three eco-epidemiological hotspots where VL has been identified are East Africa, where countries like Ethiopia, Kenya, Somalia, South Sudan, Sudan, and Uganda contribute 45% of global cases (**Figure 1** [20] and **Table 1** [21–23]) [24].

2. Main text

2.1 Distribution pattern, vector and causative agents of Leishmaniasis

Leishmaniasis is endemic in several parts of Africa. In North Africa, nations including Morocco, Algeria, Tunisia, Libya, and Egypt have recorded numerous instances of the disease, with cutaneous and visceral leishmaniasis being the most prevalent types [25]. Important species that are spread by the phlebotomus sand fly include *L. major, L. tropica, L. infantum*, and *Labrus donovani*. *L. major* is the main cause of cutaneous leishmaniasis, with *L. tropica* and *L. infantum* playing a less significant role. Visceral leishmaniasis (VL) has been linked to *L. donovani*, *L. infantum*, and infrequently *L. tropica*. Zoonotic cutaneous leishmaniasis (ZCL) and Anthropogenic cutaneous leishmaniasis (ACL) are the two kinds of cutaneous leishmaniasis seen in this region [6, 26–28]; with ZCL being caused by *L. major* and transmitted by mainly by *P. papatasi, P. duboscqi* and *P. salehi with P. papatasi* considered among the most widespread invasive sandflies species in these regions and the also has a worldwide distribution; *Psammomys obesus*, *Meriones libycus*, *Meriones shawi*, *Nesokia indica*, and *Rhombomys opimus* rodents also transmit the disease, while ACL is caused by *L. tropica* is transmitted by *P. sergenti* and gundi and hyraxes and *Leishmania killicki* with dogs also serving as accidental host in some cases [6, 29–33]. Additionally, there are two types of visceral leishmaniasis: (ZVL). Anthropogenic visceral leishmaniasis (AVL), which is caused by *L. infantum*, is primarily zoonotic, with candida serving as the principal reservoir and humans serving as unintentional hosts [34–36]. The most common form of leishmaniasis in Morocco, Algeria, and Lybia is cutaneous leishmaniasis, which is caused by the *L. tropica* species [37–40]. According to reports, *L. tropica* is a reservoir host in Ethiopia and Chad has noted an upsurge in visceral leishmaniasis in Central Africa [41, 42].

The countries of Sudan, South Sudan, Ethiopia, Kenya, Uganda, and Somalia make up one of the main geographic regions in East Africa most severely affected by VL, where the disease is primarily brought on by *Leishmania donovani* and, in certain cases, Leishmania infantum [43–45]. With the confirmation of *Phlebotomus orientalis, Phlebotomus martini, Phlebotomus celiae,* and *Phlebotomus rodhaini* as VL vectors [46–50], man and dogs serve as potential reservoir hosts. Subgenera such as Paraphlebotomus, Synphlebotomus, Larroussius, and Anaphlebotomus have also been reported [51, 52]. In East Africa, children under the age of 15 make up about 65% of VL cases, and having inadequate levels of protein, energy, iron, vitamin A, and zinc increases the risk of VL manifestation [53]. In East Africa, cutaneous and mucosal leishmaniasis have both been attributed to *Leishmania donovani* [54–56]. VL has been reported and is prevalent in Kenya, Uganda, and Somalia along their borders; *L. donovani* and *L. infantum* are the parasites responsible for the disease [57–59]. Every year, Sudan reports over two-thirds of the VL cases in East African nations; the majority of these cases are presumed to be caused by *L. donovani* [60, 61], and recent outbreaks in South Sudan have been linked to over 100,000 fatalities between 1983 and 2005 [62]. It is possible for VL to develop complications that lead to post-kala-azar cutaneous leishmaniasis, which is characterised by a macular or nodular rash that frequently starts at the mouth and is transmitted by *P. argentipes* and *P. orientalis* [63].

The disease has also been documented in Cameroon, Burkina Faso, Mauritania, Gambia, Guinea, and Nigeria in West Africa. Lieshmaniasis was originally identified in Niger, then in Nigeria, Senegal, and Mali. Mauritania, Gambia, Senegal, Nigeria, Cameroon, and Ghana are among the countries where CL is endemic and is caused by *L. major* [64, 65]. The only known case of MCL in the area is in Senegal, while the only instances of VL have been recorded in Togo and Burkina Faso with *L. donovani* as the causing agent [66]. Rodents including *Mastomys erythroleucus, Tatera gambiana, Arvicanthis niloticus*, and *Mastomys erythroleucus*, as well as dogs in Gambia, serve as the reservoir host for L. major. The parasites *L. major* and *L. donovani*, which transmit VL and CL in Ghana, Nigeria, and Senegal, respectively, are the most prevalent forms to be recorded together with *P. duboscqi* and *P. rodhaini* [66–69].

Syria has historically been endemic to CL due to *L. tropica* for centuries and, at one point, had the highest burden of this form of leishmaniasis in the region. Recent research indicates that political unrest, wars, and population migration have contributed to an increase in cases in the Middle East [70]. Due to the influx of migrants, these circumstances have caused outbreaks to be reported in 2012 in other countries including Lebanon, which have seen a rise in the number of cases recorded [71, 72]. *L. tropica* has also been claimed to be endemic in Israel and the Philistines, while the cycle of transmission is not completely understood. With a widespread geographic distribution throughout the region, *Phlebotomus (Paraphlebotomus) sergenti* has been identified as the primary vector of *L. tropica* [73–75]. *Phlebotomus arabicus* and *Phlebotomus similis* have also been documented. Leishmania major has reportedly been linked to CL in Southern Israel; this type of the illness is self-healing but makes the patient anxious. The reservoir host is the obese sand rat *Psammomys obesus* (Cricetidae: Gerbillinae), and *P. obesus* and the sand fly *Phlebotomus papatasi* (Diptera: Psychodidae) serve as the vectors [76]. *Leishmania majori*, the parasite that causes CL and VL, is the primary carrier of both the disease and its primary vector, *P. papatasi*. VL is primarily brought on by *Leishmania infantum*, though *L. tropica* can also be to blame in some cases. CL comes in two forms: ZCL and ACL, which are caused by *L. major* and *L. tropica*, respectively. Rodents are the main reservoir hosts, and some of the more endemic species include *Rhombomys* (*R.*) *opimus*, *Meriones libycus*, *Tatera* (*T.*) *indica*, and *Meriones hurrianae*. *Phlebotomus papatasi* transmits *L. major*, whereas *Phlebotomus sergenti* transmits *L. tropica* [70, 77–79]. ACL and ZCL are present, *L. tropica* and *L. major* are endemic in Saudi Arabia, and CL is declining with fewer instances being reported [80–83]. The most common kind of leishmaniasis in Iraq is CL, which is caused by the parasites *Leishmania tropica* and *Leishmania major* [84–86]. ACL and ZCL are the two types of CL that are endemic in Pakistan [87]; *L. tropica* and *L. major* are the two known forms of the parasite that transmit CL [88–90]. The frequency of ACL is higher than ZCL in Afghanistan, where *L. tropica* is known to account for 96 to 98% of CL cases [91]. ACL is said to be endemic in Turkmenistan, with *L. major* as the causative agent [8].

Countries in Europe, including Greece, have reported occasional cases of CL caused by *L. tropica*; *Phlebotomus sergenti* has been identified as the disease's primary vector and is thought to be responsible for its spread [92]. There have been reports of *Phlebotomus tobbi* and *Phlebotomus perfiliewi* as potential *L. tropica* vectors [93]. Although CL caused by *L. tropica* has not been documented in Europe, *P. sergenti* was found in Greek camps where 21 different sand fly species had significant levels of *L. tropica* infection. The emerging and re-emerging of leishmaniasis in Europe has been attributed to the introduction of exotic *Leishmania* species or strains via worldwide travelling of humans and domestic dogs; this is also responsible for the spread of visceral and cutaneous leishmaniasis caused by *L. infantum* and *L. tropicai* due to their endemicity from the Mediterranean region of Europe to neighbouring temperate areas where there are vectors without disease [90, 94]. A high risk of leishmaniasis spreading to Europe is posed by both zoonotic cutaneous and visceral leishmaniasis caused by *L. infantum* in the Mediterranean region and anthroponotic cutaneous leishmaniasis caused by *L. tropica* in Greece. Additionally, the presence of Phlebotomus sandflies without the disease is of public health concern [95]. A retrospective study of leishmaniasis across Europe revealed that cutaneous leishmaniasis was the prevalent form of the disease that occurred in 11 countries followed by visceral and mucosal disease with *L. donovani, L. infantum* (syn. *L. chagasi* in New World), *L. major, L. amazonensis* (syn. *L. garnhami*), *L. Mexicana, L. aethiopica,*

L. tropica, L. braziliensis, L. peruviana, L. guyanensis, L. panamensis, L. lainsoni, L. naiffi, L. siamensis /Lechytia martiniquensis been the parasites identified and responsible for the disease [96].

In North America, parasite species in the genus Leishmania are responsible for different clinical pathologies of the disease which include VL that is deadly and caused by *L. chagasi*, as well as MCL, localised Leishmaniasis, diffuse Leishmaniasis; MCL has been reported to be caused by *L. brasiliensis*, *L. panamensis* and *L. guyanensis*; DCL are related to *L. m. mexicana, L. amazonensis, L. pifanoi, L. guyanensis*, and *L. panamensis* [97]. Female blood-feeding sand flies of the genus Lutzomyia (Diptera: Psychodidae: Phlebotomidae) are the transmission vector in Mexico, the United States, and Canada; humans are incidental hosts while mammals serve as reservoir hosts; *L. mexicana* is responsible for the majority of human cases that have been reported [98]. Leishmania infantum, which causes canine visceral leishmaniasis, has been reported in dogs and cats. While cutaneous *Leishmania mexicana* is endemic to the United States, visceral leishmaniasis is not considered endemic among humans but primarily affects dogs, which are thought to be the reservoir host for humans [99, 100].

Leishmania (Viannia) braziliensis, L. (Viannia) amazonensis, L. (Viannia) guyanensis, and *L. Infantum*, associated with visceral leishmaniasis, were the four species of Leishmania cases reported in humans in South America. In addition, 28 species of Phlebotominae (23 species of Lutzomyia, 4 species of Brumptomyia, and 1 species of [101]. Both CL and MCL are induced by *L. Braziliensis*: *Lutzomyia neivai*, *L. Whitmani*, *L. Cortelezzii* complex (*L. Cortelezzii, L. Sallesi*), *L. Migonei*, and *L. Pessoai*, while V is brought on by *L. Infantum* was *L. Longipalpis* [102]. However, domestic and wild canids have been well established as the reservoir host of leishmania, with more than 20 species of the parasites being transmitted by the phlebotomine vectors (Diptera, Psychodidae); CL and VL are the two major forms of the disease where *Leishmania (Viannia) braziliensis*, *Leishmania (Leishmania) amazonensis, L. mexicana, L. donovani* causes VL in the old world, but *L. infantum* does so in the new world [103–107]. In Colombia several species of *Leishmania* parasites have been reported to infect humans with CL been the major form of the disease followed by MCL and VL; *L. braziliensis*, *L. panamensis* and *L. guyanensis* have been responsible for the outbreaks of CL; other parasites include *L. lainsoni*, *L. amazonensis*, *L. infantum chagasi*, *L. Mexicana*, *L. colombiensis* and *L. equatoriensis,* this country is ranked second only to Brazil in terms of cases of leishmaniasis [108–110]. In Bolivia, *Leishmania (Viannia) lainsoni, Leishmania (Viannia) guyanensis, Leishmania (Viannia) panamensis and Leishmania (Leishmania) amazonensis* have been described as a causal agent of Tegumentary leishmaniasis (TL) (CL, MCL, DCL) with transmission also reported to be zoonotic and caused by *Nyssomyia neivai,* Cortelezzii complex, *Evandromyia sallesi*, *Migonemyia migonei* and *Micropygomyia quinquefer* [111].

The parasite *L. donovani*, the vector *Phlebotomus annandalei*, Phlebotomus argentipes sensu stricto, and Phlebotomus glaucus, and the disease CL, MCL, and VL are all endemic throughout Asia [112]. VL, which is indigenous to Nepal and brought on by *L. donovani*, can occasionally lead to post-Kala-azar cutaneous leishmaniasis [113]. Similar to Kala-azar sickness, VL is chronic in Bangladesh and can lead to it. *L. donovani* is thought to be the main cause of the illness, while *Phlebotomus papatasi* is the disease's vector in Asia [114, 115].

Leishmania donovani is the chronic disease that causes VL on the Indian subcontinent; however, *L. tropica* or *L. infantum*, particularly when co-infected with HIV, can also cause the illness; *Phlebotomus argentipes* is the vector for VL, with humans

Clinical forms	Leishmania parasite	Geographical distribution
CL	*L. major*, *L. tropica*, *L. infantum*, and *Labrus donovani; L. braziliensis*: *Lutzomyia neivai*, *Lu. Whitmani*, *L. Cortelezzii complex* (*L. Cortelezzii–L. Sallesi*), *L. Migonei*, and *L. Pessoai*, *L. (L.) peruviana*, *L. lainsoni*, *L. amazonensis*, *L. infantum chagasi*, *L. Mexicana*, *Leptodactylus colombiensis* and *L. equatoriensis*	Morocco, Algeria, Tunisia, Libya, Egypt; Chad; Sudan, South Sudan, Ethiopia, Kenya, Uganda and Somalia; Niger, Nigeria, Senegal and Mali; Cameroon, Burkina Faso, Mauritania, Gambia and Guinea, Ghana; Syria, Lebanon, Israel, Philistines, Iran, Saudi Arabia, Iraq, Pakistan, Afghanistan, Turkmenistan; Greece; United State of America; Bolivia, Brazil, Colombia, Peru; Sri Lanka, China, Nepal Bangladish; India, and
MCL	*L. viannia braziliensis, L. viannia amazonesis, L. viannia panamensis* and *L viannia guyanesis*	Senegal; Colombia; Sri Lanka
DCL	*L. Mexicana, L. aethiopica, L. amazonensis, L. pifanoi, L. guyanensis*, and *L. panamensis*	Mexico, United States, and Canada
VL	*L. donovani, L. infantum, L. tropica, L. chagasi*	Ethiopia, Kenya, Somalia, South Sudan, Sudan, Chad, Togo, Burkina Faso, Cameroon and Uganda; Iran, Bangladesh, Sri Lanka and Nepal; India; Brazil, Argentina, Colombia; India

Table 2.
The table below shows the clinical form of leishmaniasis and their geographical distribution [21–23].

functioning as reservoir hosts. PKDL does not have an animal reservoir, and transmission is thought to be anthroponotic [116, 117]. VL is also known as kala-azar, which is the Hindi word for "black disease" (**Table 2**).

3. Leishmaniasis and HIV co-infection

There have been reports of leishmania-HIV co-infection in 45 countries in 2021, with Brazil, Ethiopia, and Bihar in India having the highest number of cases [118]. According to a report by the WHO in 2022, people who have leishmaniasis and HIV are at an increased risk of developing full clinical disease, high relapse rates, and mortality rates. Leishmaniasis and HIV co-infection have also been shown to occur in CL and VL [119, 120]. A change in immune response brought on by immunosuppression caused by a low CD4 T-lymphocyte count and the development of immunological activation with cell senescence are two examples of the clinical manifestation of VL [121].

4. Risk factors in the distribution of Leishmaniasis

Conflict, climate (temperature, rainfall, relative humidity, precipitation), physical (geographical barriers), and biotic (distribution and abundance of vertebrate hosts) factors, socioeconomic circumstances, environmental (land degradation, agricultural activities, deforestation due to urbanisation), domestic animals and living standards, immune response of hosts, and conflict have all been linked to the distribution of leishmaniasis [6]. Although no consistent pattern has been identified, civil wars, social unrest, migration, forced population displacement, rainfall, humidity, temperature, soil types and moisture contents, and land cover type are significantly associated with the distribution pattern of these sandflies and contribute to emerging and re-emerging outbreaks [18].

5. Climatic conditions

Leishmaniasis is a disease that is sensitive to climate because variations in temperature, precipitation, and humidity have a significant impact on the life cycles of the parasite's vectors and reservoir hosts. Climate changes can also affect the distribution, survival, and population of the vectors and reservoir hosts [122]. Sandflies have been reported to be abundant between the months of June and September with *P. argentipes* being active between this period in India with temperature ranging between 27 and 31°C [123, 124]. Fluctuations in temperature can negatively affect the developmental life cycle of leishmania promastigote in sandflies, which can result in the transmission of the parasite into areas not previously endemic for the disease.

5.1 Vegetation cover

Leishmaniasis outbreaks have been shown to decrease with the improvement of vegetation types that limit rodent growth [125]; it has also been established that areas with thick or dense vegetation cover have a higher risk of leishmaniasis infection and transmission [126]. Forest fragmentation and deforestation have been linked to an increased risk for leishmaniasis.

5.2 Age and gender

Leishmaniasis has been found to affect adults more frequently than children; this could be because younger people are less exposed to the parasites and vectors that cause the disease, especially in rural areas where adults are more engaged in farming and other socioeconomic activities than children are. It has also been found that men are generally more susceptible to infection, though this could vary [127, 128].

5.3 Housing and socio-economic conditions

Poor housing and the lack of good sanitary conditions in terms of waste management, sewage disposal, and drainages in the majority of rural and semi-urban centre provide breeding and resting sites for sandflies, bringing this vector closer to human habitation. These sandflies feed on blood meal. Other good breeding sites for these vectors include cracks and crevices in walls, roofs that are poorly made, and thatched homes [129]. The disease spreads as a result of people moving from rural to urban areas for economic reasons [130–132].

5.4 Domestic animal breeding

It has been discovered that there may be a significant risk of infection in houses where animals are housed, bred, or live in close proximity to people. This is because some of these animals may act as reservoir hosts for the disease; leishmaniasis, particularly the VL, affects dogs and felines [133].

5.5 Host immune responses and parasite factor

In order for the infection to take hold, promastigotes must infiltrate macrophages and avoid inducing host defences. Additionally, the progression of intracellular infection by amastigotes depends on the preservation of macrophages in an inert, deactivated state. Through cell-mediated immunity and delayed-type hypersensitivity, an immunocompetent host defence system is equipped with and reacts to non-specific innate and antigen-specific acquired mechanisms; these inflammatory responses play a critical role in disease expression and may result in asymptomatic infection and self-healing or may result in clinical manifestation of disease. People with weakened immune systems, such as HIV patients, are more susceptible to infection since their immune systems are unable to defend them against the parasites. The parasite, however, can alter intracellular kinases, phosphatases, signalling pathways, the responsiveness of macrophages to the parasite and secretion of cytokines, as well as dendritic cell and macrophage mechanisms. This allows the parasites in the case of an infection to alter, temper, or hinder with intracellular kinases, phosphatases, signalling pathways, the responsiveness of macrophages to the parasite and secretion of the parasite can also disrupt dendritic cells, which are important for antigen presentation, T-cell co-stimulation, and the effective establishment of acquired Th1 responses [134].

6. Life cycle of leishmania parasite

Leishmania parasites go through two main stages in their life cycle: the extracellular stage in invertebrate hosts and the intracellular stage in vertebrate hosts. The parasite has two primary morphologies: amastigote and promastigote, with the former occurring in vertebrate hosts and the latter in invertebrate hosts. The promastigotes are injected into the host circulation by female sandflies during a blood meal on a vertebrate host; after this inoculation, the promastigotes are phagocytosed by the host macrophages; the macrophage ruptures and releases a large number of amastigotes into circulation. Monocytes and macrophages in the spleen, liver, lymph nodes, bone marrow, and other tissues of reticulo-endothelial cells are invaded by the released amastigotes. The female sandflies then consume free amastigotes from the blood as well as intracellular amastigotes from the monocytes. The promastigotes in the sand fly's midgut move over a period of 6 to 9 days to its pharynx and buccal cavity, where the sand fly can spread the parasites to a new host after a blood meal [135].

7. The vector for leishmaniasis

Only 21 of the 700 species of sandflies (family Phlebotominae) are known to be disease vectors. The majority of these species belong to the genera Phlebotomus, Sergentomyia, and Lutzomyia; the Lutzomyia group includes species from the old and new worlds. Old-world species are those that can be found in South and Central Asia, the Indian subcontinent, the Middle East, North and East Africa, and Southern Europe; new-world species are those that can be found in Brazil and other Latin American nations, Mexico, and the United States. These fly species are absent from Australia but are indigenous to over 90 other nations and are found on every continent [136]. The following list includes the phlebotomous sandfly species that cause leishmaniasis in the old world. The transmission of *L. major, L. tropica, L. infantum,* and *L. aethiopica* is carried out by *P. longicuspis, P. papatasi, P. ariasi, P. pedifer, P. duboscqi, P. rhodaini, P. segenti, and P. sergenti.* The new world sandfly genus Lutzomyia includes species including *L. hartmanni*, *L. gomezi*, *L. longipalpis, L. Ayacuchensis, L. Sanguinaria, L. Trapidoire* responsible for the transmission of *L. infantum, L. vianna equatorensis, L. colombiensis*, and *L. vianna braziliensis*, among others. The Psychodopygus, which includes species like *P. panamensis, P. carrerai*, and *P. amazonensis*, is another type of sandfly of significant medical and veterinary significance [42, 134–144].

8. Conclusions

L. donovani, *L. tropica*, and *L. major* are the species that cause infection in Africa, L. tropica, L. major, and Leishmania infantum are the species that cause infection in the Middle East, *L. tropica, L. infantum*, and Europe, *L. mexicana* and *L. chagasi* are the species that cause infection in North America, and *Leishmania (Viannia) braziliensis* is the species that causes infection in South America. The disease is spread by numerous different species of Phlebotomus sandflies. Risk factors for the disease include climatic factors, vegetation cover, age and gender, housing and socioeconomic circumstances, animal breeding, and host immunity. Additionally, migration of people from other continents has been to blame for the emergence and re-emergence of the disease in Europe where there is civil and regional strife, political instability, and increased disease transmission and dissemination between continents. Obstacles to the quest to eradicate the disease include limited access to health care, a lack of national policy, and the maintenance of such policy, if it exists, particularly in developing nations.

Acknowledgements

Non to be disclosed.

Conflict of interest

The authors declare no conflict of interest.

DOI: http://dx.doi.org/10.5772/intechopen.110490

Author details

Stephen Kyari
Department of Veterinary Parasitology and Entomology, Ahmadu Bello University, Zaria, Kaduna State, Nigeria

*Address all correspondence to: kyaristephen757@gmail.com

References

[1] Ahmed K, Abdelkrim BS, Denis S, Samir, El J, Lhoussain H. Geographic distribution of *Meriones shawi, Psammomys obesus,* and *Phlebotomus papatasi the main reservoirs and principal vector* of zoonotic cutaneous leishmaniasis in the Middle East and North Africa. Parasite Epidemiology and Control. 2022;**17**:e00247

[2] Javadian E, Dehestani M, Nadim A, Rassi Y, Tahvildar-Bidruni G, Seyedi-Rashti M, et al. Confirmation of *Tatera indica* (Rodentia: Gerbilldae) as the main reservoiur host of zoonotic cutaneous leishmaniasis in the west of Iran. Iranian Journal of Public Health. 1998;**27**(1-2):55-60

[3] Rasi Y, Jalali M, Javadian E, Motazedian MH. Confirmation of *Meriones libycus* (Rodentia; Gerbillidae) as the main reservoiur host of zoonotic cutaneous leishmaniasis in Arsanjan, Fars Province, south of Iran (1999-2000). Iranian Journal of Public Health. 2001;**30**(3-4):143-144

[4] Afshar AA, Rassi Y, Sharifi I, Abai M, Oshaghi M, Yaghoobi-Ershadi M, et al. Susceptibility status of Phlebotomus papatasi and P. sergenti (Diptera: Psycholididae) to DDT and deltamethrin in a focus of cutaneous leishmaniasis after earth quake strike in bam, Iran. Iranian Journal of Arthropod-Borne Diseases. 2011;**5**(2):32 PMCID: PMC3385580. PMID: 22808416

[5] Gholamrezaei M, Mohebali M, Hanaf-Bojd AA, Sedaghat MM, Shirzadi MR. Ecological niche modeling of main reservoir hosts of zoonotic cutaneous leishmaniasis in Iran. Acta Tropica. 2016;**160**:44-52. DOI: 10.1016/j.actatropica.2016.04.014

[6] Tabbabi A. Review of Leishmaniasis in the Middle East and North Africa. African Health Sciences. 2019;**19**(1):1329-1337. DOI: 10.4314/ahsv19i1.4

[7] WHO expert committee on the control of the leishmania and World Health Organization (2010) control of leishmaniases: Report of a meeting of the WHO expert committee on the control of leishmaniases Geneva, 22-26 March 2010

[8] Alvar J, Vélez ID, Bern C, Herrero M, Desjeux P, Cano J. Leishmaniasis worldwide and global estimates of its incidence. PLoS One. 2012;7:e35671. DOI: 10.1371/journal.pone.0035671

[9] Hafza I, Muhammad ZK. Types and treatments of Leishmaniasis. Journal of Scientific and Technical Research. 2022;**42**(5):34037-34042. DOI: 10.26717/BJSTR.2022.42.006815

[10] Tonio VP, Charles MA. Leishmaniasis. Postgraduate Medical Journal. Feb 2007;**83**(976):649-657. DOI: 10.1136/pgmj.2006.047340corr1

[11] Marsden PD, Nonata RR. Mucocutaneous leishmaniasis—A review of clinical aspects. Revista da Sociedade Brasileira de Medicina Tropical. 1975;**9**:309-326

[12] Abadías-Granado I, Diago A, Cerro PA, Palma-Ruiz AM, Gilaberte Y. Cutaneous and Mucocutaneous Leishmaniasis. Actas Dermo-Sifiliográficas. 2021;**112**:601-618

[13] Ahluwalia S, Lawn SD, Kanagalingam J, Grant H, Lockwood DN. Mucocutaneous leishmaniasis: An imported infection among travellers to central and South America. BMJ. 2004;**329**:842-844

[14] David CV, Craft N. Cutaneous and mucocutaneous leishmaniasis. Dermatologic Therapy. 2009;**22**(6):491-502

[15] David CR, Llanos-Cuentas EA, Campos P, et al. Spraying houses in the Peruvian Andes with lambda-cyhalothrin protects residents against cutaneous leishmaniasis. Transactions of the Royal Society of Tropical Medicine and Hygiene. 2012;**94**:631-636

[16] Mauel J. Vaccination against *Leishmania* infections. Current Drug Targets - Immune, Endocrine & Metabolic Disorders. 2002;**2**:201-226

[17] Manson-Bahr PEC, Apted FIC. Leishmaniasis. In: Manson-Bahr PEC, Apted FIC, editors. Manson's Tropical Diseases. 18th ed. London: Bailliere Tindall; 1982. pp. 93-115

[18] Zijlstra EE, El-Hassan AM. Visceral leishmaniasis. Transactions of the Royal Society of Tropical Medicine and Hygiene. 2001;**95**(1):S27-S58

[19] Wilson ME, Streit JA. Visceral leishmaniasis. Gastroenterology Clinics of North America. 1996;**25**:535-551

[20] Esteves S, Costa I, Amorim C, Santarem N, Cordeiro-da-Silva A. Biomarkers in leishmaniasis: From basic research to clinical application [Internet]. In: Biomarker - Indicator of Abnormal Physiological Process. InTech; 2018. DOI: 10.5772/intechopen.75315

[21] Sosa N, Pascale JM, Jiménez AI, Norwood JA, Kreishman-Detrick M, Weina PJ, et al. Topical paromomycin for New World cutaneous leishmaniasis. PLoS Neglected Tropical Diseases. 2019;**13**(5):e0007253. DOI: 10.1371/journal.pntd.0007253

[22] Mann S, Phupitakphol T, Davis B, Newman S, Suarez JA, Henao-Martínez A, et al. Case report: Cutaneous Leishmaniasis due to Leishmania (Viannia) panamensis in two Travelers successfully treated with Miltefosine. The American Journal of Tropical Medicine and Hygiene. 2020;**103**(3):1081-1084. DOI: 10.4269/ajtmh.20-0086

[23] Madusanka RK, Hermali S, Nadira DK. Treatment of cutaneous Leishmaniasis and insights into species-specific responses: A narrative review. Infectious Disease and Therapy. 2022;**11**:695-711. DOI: 10.1007/s40121-022-00602-2

[24] World Health Organization (WHO). Regional Office for Europe. Vector Borne and Rodent Borne Diseases. WHO; 2009. Available from: http://www.euro.who.int/globalchange/Assessment/20070216_10

[25] McDowell MA, Rafati S, Ramalho-Ortigao M, Ben SA. Leishmaniasis: Middle East and North Africa research and development priorities. PLoS Neglected Tropical Diseases. 2011;**5**:e1219. DOI: 10.1371/journal.pntd.0001219

[26] Tabbabi A, Ghrab J, Aoun K, Ready PD, Bouratbine A. Habitats of the sandfly vectors of *Leishmania tropica* and *L. major* in a mixed focus of cutaneous leishmaniasis in south East Tunisia. Acta Tropica. 2011;**119**:131-137. DOI: 10.1016/j.actatropica.2011.05.002

[27] Tabbabi A, Bousslimi N, Rhim A, Aoun K, Bouratbine A. First report on natural infection of Phlebotomus sergenti with *Leishmania promastigotes* in the cutaneous leishmaniasis focus in SouthEastern Tunisia. American Journal of Tropical Medicine and Hygiene. 2011;**85**:646-647. DOI: 10.4269/ajtmh.2011.10-0681

[28] Postigo JA. *Leishmaniasis* in the World Health Organization eastern Mediterranean region. International Journal of Antimicrobial Agents. 2010;**36**(Suppl. 1):S62-S65. DOI: 10.1016/j.ijanti micag.2010.06.023

[29] Hamadto HA, Al FA, Farrag AB, Abdel Maksoud MK, Morsy TA. Zoonotic cutaneous leishmaniasis: Reservoir host and insect vector in North Sinai, Egypt. Journal of the Egyptian Society of Parasitology. 2007;**37**:843-850

[30] Chelbi I, Kaabi B, Béjaoui M, Derbali M, Zhioua E. Spatial correlation between *Phlebotomus papatasi Scopoli* (Diptera: Psychodidae) and incidence of zoonotic cutaneous *leishmaniasis* in Tunisia. Journal of Medical Entomology. 2009;**46**:400-402 PubMed

[31] Harrat Z, Boubidi SC, Pratlong F, Benikhlef R, Selt B, et al. Description of a dermatropic *leishmania* close to *L. killicki* (Rioux, Lanotte & Pratlong in Algeria). Transactions of the Royal Society of Tropical Medicine and Hygiene. 2009;**103**:716-720. DOI: 10.1016/j.trstmh.2009.04.013

[32] Fathy FM, El-Kasah F, El-Ahwal AM. Emerging cutaneous leishmaniasis in Sirte-Libya: Epidemiology, recognition and management. Journal of the Egyptian Society of Parasitology. 2009;**39**:881905

[33] Lemrani M, Nejjar R, Pratlong F. A new *Leishmania tropica* zymodeme—Causative agent of canine visceral leishmaniasis in northern Morocco. Annals of Tropical Medicine & Parasitology. 2002;**96**:637-638. DOI: 10.1179/000349802125001645 PMID: 12396327

[34] Aoun K, Amri F, Chouihi E, et al. *killicki* en Tunisie: Résultats et analyse de l'identifcation de 226 isolats humains et canins et revue de la littérature. Bulletin de la Societe de Pathologie Exotique. 2008;**101**:323-328

[35] Aoun K, Jeddi F, Amri F, et al. Actualités épidémiologiques de la *leishmaniose* viscérale en Tunisie. Médecine et Maladies Infectieuses. 2009;**39**:775-779. DOI: 10.1016/j.medmal.2009.08.010

[36] Ghrab J, Rhim A, Bach HD, et al. Phlebotominae (*Diptera:Psychodidae*) of human leishmaniosis sites in Tunisia. Parasite. 2006;**13**:23-33. DOI: 10.1051/parasite/2006131023

[37] El Idrissi SI, Benlabsir C, Fellah H, Lemrani M, Riyad M. Transmission patterns of Leishmania tropica around the Mediterranean basin: Could Morocco be impacted by a zoonotic spillover? PLOS Neglected Tropical Diseases. 2022;**16**(1):e0010009. DOI: 10.1371/journal.pntd.0010009

[38] Kbaich MA, Mhaidi I, Ezzahidi A, Dersi N, El Hamouchi A, Riyad M, et al. New epidemiological pattern of cutaneous leishmaniasis in two pre-Saharan arid provinces, southern Morocco. Acta Tropica. 2017;**173**:11-16. DOI: 10.1016/j.actatropica.2017.05.016. PMID: 28527882

[39] Chaara D, Haouas N, Dedet JP, Babba H, Pratlong F. Leishmaniases in Maghreb: An endemic neglected disease. Acta Tropica. 2014;**132**:80-93. DOI: 10.1016/j.actatropica.2013.12.018 PMID:24412727

[40] Eddaikra N, Ait-Oudhia K, Kherrachi I, Oury B, Moulti-Mati F, Benikhlef R, et al. Antimony susceptibility of Leishmania isolates collected over a 30-year period in Algeria. PLoS Neglected Tropical Diseases. 2018;**12**(3):e0006310. DOI: 10.1371/journal.pntd.0006310. PMID: 29561842

[41] Kassahun A, Sadlova J, Benda P, Kostalova T, Warburg A, Hailu A, et al. Natural infection of bats with Leishmania in Ethiopia. Acta Tropica. 2015;**150**:166-170. DOI: 10.1016/j.actatropica.2015.07.024 PMID: 26232657

[42] Israël DK, Gonnonta V, Gunter MC, Coulibaliy A, Seydou D. Retrospective study of visceral Leishmaniasis in the Health District of Léré in Chad, Central Africa. Research Square. 2021:1-9. DOI: 10.21203/rs.3.rs-401673/v1

[43] Malaria Consortium. Leishmaniasis control in eastern Africa: Past and present efforts and future needs. Situation and gap analysis. Malaria Consortium. 2010;**87**. Available from: https://assets.publishing.service.gov.uk/media/57a08b0ae5274a31e0000920/Leishmaniasis_Eastern_Africa_Situation_Analysis.pdf

[44] Alvar J, Bashaye S, Argaw D, Cruz I, Aparicio P, et al. Kala-azaroutbreak in Libo Kemkem, Ethiopia: Epidemiologic and parasitologic assessment. The American Journal of Tropical Medicine and Hygiene. 2007;**77**:275-282

[45] Leta S, Dao THT, Mesele F, Alemayehu G. Visceral Leishmaniasis in Ethiopia: An evolving disease. PLoS Neglected Tropical Diseases. 2014;**8**(9):e3131. DOI: 10.1371/journal.pntd.0003131

[46] Dawit G, Girma Z, Simenew K. A review on biology, epidemiology and public health significance of Leishmaniasis. Journal of Bacteriology & Parasitology. 2013;**4**:166

[47] CDC. Parasites - Leishmaniasis. Centers for Disease Control and Prevention. 2013. Available from: http://www.cdc.gov/parasites/leishmaniasis/ [Accessed: 6 August 2014]

[48] Gebre-Michael T, Lane R. The roles of Phlebotomus martini and P.celiae (Diptera: Phlebotominae) as vectors of visceral leishmaniasis in the aba Roba focus, southern Ethiopia. Medical and Veterinary Entomology. 1996;**10**:53-62

[49] Gebre-Michael T, Malone JB, Belkew M, Ali A, Berhe N, et al. Mapping the potential distribution of Phlebotomus martini and P.orientalis (Diptera: Psychodidae), vectors of kala-azar in East Africa by use of geographic information systems. Acta Tropica. 2004;**90**:73-86

[50] Gebre-Michael T, Balkew M, Berhe N, Hailu A, Mekonnen Y. Further studies on the phlebotomine sandflies of the kala-azar endemic lowlands of Humera-Metema (north-West Ethiopia) with observations on their natural blood meal sources. Parasites & Vectors. 2010;**3**:6

[51] Anjili CO, Ngumbi PM, Kaburi JC, Irungu LW. The phlebotomine sandfly fauna (Diptera: Psychodidae) of Kenya. Journal of Vector Borne Diseases. 2011;**48**(4):183-189

[52] Al-Salem W, Jennifer RH, Peter JH. A review of visceral leishmaniasis during the conflict in South Sudan and the consequences for east African countries. Parasites & Vectors. 2016;**9**:460. DOI: 10.1186/s13071-016-1743-7

[53] WHO. Dramatic Upsurge in Visceral Leishmaniasis Cases in the Horn of Africa. Press release; 1998. Available from: http://www.who.int/inf-pr-1998/en/pr98-23.html

[54] Zijlstra EE, Khalil EA, Kager PA, El-Hassan AM. Post-kala-azar dermal leishmaniasis in the Sudan: Clinical presentation and differential diagnosis. The British Journal of Dermatology. 2000;**143**(1):136-143

[55] Dereure J, El-Safi SH, Bucheton B, Boni M, Kheir MM, Davoust B, et al. Visceral leishmaniasis in eastern Sudan: Parasite identification in humans and dogs; host-parasite relationships. Microbes and Infection. 2003;**5**(12):1103-1108

[56] Mahdi M, Elamin EM, Melville SE, Musa AM, Blackwell JM, Mukhtar MM, et al. Sudanese mucosal leishmaniasis: Isolation of a parasite within the Leishmania donovani complex that differs genotypically from L. donovani causing classical visceral leishmaniasis. Infection, Genetics and Evolution. 2005;**5**(1):29-33

[57] Mugasa CM, Deborggraeve S, Schoone GJ, Laurent T, Leeflang MM, Ekangu RA, et al. Accordance and concordance of PCR and NASBA followed by oligochromatography for the molecular diagnosis of Trypanosoma brucei and Leishmania. Tropical Medicine & International Health. 2010;**15**(7):800-805

[58] Raguenaud ME, Jansson A, Vanlerberghe V, Deborggraeve S, Dujardin JC, Orfanos G, et al. Epidemiology and clinical features of patients with visceral leishmaniasis treated by an MSF clinic in Bakool region, Somalia, 2004-2006. PLoS Neglected Tropical Diseases. 2007;**1**(1):e85

[59] Marlet MV, Wuillaume F, Jacquet D, Quispe KW, Dujardin JC, Boelaert M. A neglected disease of humans: A new focus of visceral leishmaniasis in Bakool, Somalia. Transactions of the Royal Society of Tropical Medicine and Hygiene. 2003;**97**(6):667-671

[60] Abbasi I, Aramin S, Hailu A, Shiferaw W, Kassahun A, Belay S, et al. Evaluation of PCR procedures for detecting and quantifying Leishmania donovani DNA in large numbers of dried human blood samples from a visceral leishmaniasis focus in northern Ethiopia. BMC Infectious Diseases. 2013;**13**:153

[61] Gelanew T, Kuhls K, Hurissa Z, Weldegebreal T, Hailu W, Kassahun A, et al. Inference of population structure of Leishmania donovani strains isolated from different Ethiopian visceral leishmaniasis endemic areas. PLoS Neglected Tropical Diseases. 2010;**4**(11):e889

[62] Wakabi W. Peace has come to southern Sudan, but challenges remain. Lancet. 2006;**368**(9538):829-830

[63] Desjeux P, Ghosh RS, Dhalaria P, Strub-Wourgaft N, Zijlstra EE. Report of the post kala-azar dermal leishmaniasis (PKDL) consortium meeting, New Delhi, India, 27-29 June 2012. Parasites & Vectors. 2012;**6**:196

[64] Boakye DA, Wilson MD, Kweku M. A review of leishmaniasis in West Africa. Ghana Medical Journal. 2006;**39**:94-97

[65] Kimutai A, Peter KN, Willy Kiprotich T, Michael MG, Lydia BN. Leishmaniasis in northern and western Africa: A review. African Journal of Infectious Diseases. 2009;**3**(1):14-25

[66] Donji B, Dereure J, Poste B, Same-Ekobo A, Dedet JP. Visceral leishmaniasis in Cameroon. Sero-epidemiological survey in the Kousseri region, North Cameroon. Bulletin de la Societe de Pathologie Exotique. 2001;**94**:418-420

[67] Thierno DM, Develoux M, Ndiaye B, Huerre M. Infiltrated cutaneous leishmaniasis and sporotrichosis caused by *Leishmania major*. First Senegalese case. Bulletin de la Societe de Pathologie Exotique. 2001;**94**:19-20

[68] Fryauff JD, Hanafi AH, Klena JD, Hoel DF, Appawu M, Rogers W, et al. Short report: ITS-1 DNA sequence confirmation of *Leishmania major* as a cause of cutaneous leishmaniasis from an outbreak focus in the Ho district, Southeastern Ghana. The American Journal of Tropical Medicine and Hygiene. 2006;**75**:502-504

[69] Niamba P, Goumbri-Lompo O, Traoré A, Barro-Traoré F, Soudré RT. Diffuse cutaneous leishmaniasis in an HIV-positive patient in western Africa. The Australasian Journal of Dermatology. 2007;**48**:32-34

[70] Ghatee MA, Taylor WR, Karamian M. The geographical distribution of cutaneous Leishmaniasis causative agents in Iran and its Neighboring countries, a review. Frontiers in Public Health. 2020;**18**:8-11. DOI: 10.3389/fpubh.2020.00011 PMID: 32133334

[71] Alawieh A, Musharrafieh U, Jaber A, Berry A, Ghosn N, Bizri AR. Revisiting leishmaniasis in the time of war: The Syrian conflict and the Lebanese outbreak. International Journal of Infectious Diseases. 2014;**29**:115-119. https://doi.org/10.1016/j.ijid.2014.04.023 PMID: 25449245

[72] Saroufim M, Charafeddine K, Issa G, Khalifeh H, Habib RH, Berry A, et al. Ongoing epidemic of cutaneous leishmaniasis among Syrian refugees, Lebanon. Emerging Infectious Diseases. 2014;**20**:1712-1715. DOI: 10.3201/eid2010.140288 PMID: 25279543

[73] Kamhawi S, Modi GB, Pimenta PFP, Rowton E, Sacks DL. The vectorial competence of Phlebotomus sergenti is specific for *Leishmania tropica* and is controlled by species-specific, lipophosphoglycanmediated midgut attachment. Parasitology. 2000;**121**:25-33. DOI: 10.1017/s0031182099006125 PMID:11085222

[74] Depaquit J, Ferté H, Léger N, Lefranc F, Alves-Pires C, Hanafi H, et al. ITS2 sequences heterogeneity in Phlebotomus sergenti and Phlebotomus similis (Diptera, Psychodidae): Possible consequences in their ability to transmit Leishmania tropica. International Journal for Parasitology. 2002;**32**:1123-1131. DOI: 10.1016/s0020-7519(02)00088-7 PMID:12117495

[75] Maroli M, Feliciangeli MD, Bichaud L, Charrel RN, Gradoni L. Phlebotomine sandflies and the spreading of leishmaniases and other diseases of public health concern. Medical and Veterinary Entomology. 2013;**27**:123-147. DOI: 10.1111/j.1365-2915.2012.01034.x PMID: 22924419

[76] Wasserberg G, Abramsky Z, Kotler BP, Ostfeld RS, Yarom I, Warburg A. Anthropogenic disturbances enhance occurrence of cutaneous leishmaniasis in Israel deserts: Patterns and mechanisms. Ecological Applications. 2003;**13**(3):868-881

[77] Shiravand B, Hanafi-Bojd AA, Tafti AAD, Abai MR, Almodarresi A, Mirzaei M. Climate change and potential distribution of zoonotic cutaneous leishmaniasis in Central Iran: Horizon 2030 and 2050. Asian Pacific Journal of Tropical Medicine. 2019;**12**(5):204-215

[78] Schnur LF, Nasereddin A, Eisenberger CL, Jaffe CL, El Fari M, Azmi K, et al. Multifarious characterization of *Leishmania tropica* from a Judean desert focus, exposing intraspecific diversity and incriminating *Phlebotomus sergenti* as its vector. The American Journal of Tropical Medicine and Hygiene. 2004;**70**:364-372. DOI: 10.4269/ajtmh.2004.70.364

[79] Farash BRH, Shamsian SAA, Mohajery M, Fata A, Sadabadi F, Berenji F, et al. Changes in the epidemiology of cutaneous Leishmaniasis in northern Iran. Türkiye Parazitolojii Dergisi. 2020;**44**(1):52-57

[80] Mondragon-Shem K, Al-Salem WS, Kelly-Hope L, Abdeladhim M, Al-Zahrani MH, Valenzuela JG, et al. Severity of old world cutaneous leishmaniasis is influenced by previous exposure to sandfly bites in Saudi Arabia. PLoS Neglected Tropical Diseases. 2015;**9**:e0003449. DOI: 10.1371/journal.pntd.0003449

[81] El-Beshbishy HA, Al-Ali KH, El-Badry AA. Molecular characterization of cutaneous leishmaniasis in Al-Madinah Al-Munawarah province, western Saudi Arabia. International Journal of Infectious Diseases. 2013;**17**:e334-e338. DOI: 10.1016/j.ijid.2012.11.015 92

[82] Haouas N, Amer O, Alshammri FF, Al-Shammari S, Remadi L, Ashankyty I. Cutaneous leishmaniasis in northwestern Saudi Arabia: Identification of sand fly fauna and parasites. Parasites & Vectors. 2017;**10**:544. DOI: 10.1186/s13071-017-2497-6

[83] Zakai HA. Cutaneous leishmaniasis in Saudi Arabia: current status. Journal of Advanced Laboratory Research in Biology. 2014;**5**:29-34

[84] Salam N, Al-Shaqha WM, Azzi A. Leishmaniasis in the Middle East: Incidence and epidemiology. PLoS Neglected Tropical Diseases. 2014;**8**:e3208. DOI: 10.1371/journal.pntd.0003208

[85] Ali MA, Rahi AA, Khamesipour A. Species typing with PCR–RFLP from cutaneous leishmaniasis patients in Iraq. Donnish Journal of Medicine and Medical Sciences. 2015;**2**:026-031

[86] Al-Bajalan MM, Al-Jaf SM, Niranji SS, Abdulkareem DR, Al-Kayali KK, Kato H. An outbreak of *Leishmania major* from an endemic to a non-endemic region posed a public health threat in Iraq from 2014-2017: Epidemiological, molecular and phylogenetic studies. PLoS Neglected Tropical Diseases. 2018;**12**:e0006255. DOI: 10.1371/journal.pntd.0006255

[87] Hajra H, Nazar-ul I, Khalid M, Sajid A. Occurrence analysis of cutaneous Leishmaniasis particularly in Pakistan. Biomedical Journal of Scientific & Technical Research. 2019;**19**(5):1-5. BJSTR.MS.ID.003359

[88] Hussain M, Munir S, Ayaz S, Khattak BU, Khan TA, Muhammad N, et al. First report on molecular characterization of *Leishmania* species from cutaneous leishmaniasis patients in southern Khyber Pakhtunkhwa province of Pakistan. Asian Pacific Journal of Tropical Medicine. 2017;**10**:718-721. DOI: 10.1016/j.apjtm.2017.07.015

[89] Ghatee MA, Sharifi I, Kuhls K, Kanannejad Z, Harandi MF, de Almeida ME, et al. Heterogeneity of the internal transcribed spacer region in *Leishmania tropica* isolates from southern Iran. Experimental Parasitology. 2014;**144**:44-51. DOI: 10.1016/j.exppara.2014.06.003

[90] Naseer U, Muzafar S, Mian SKA. Brief review on infestation of cutaneous Leishmaniasis in Pakistan. Biomedical Journal of Scientific & Technical Research. 2020;**31**(4):1-4. BJSTR.MS.ID.005141

[91] Fakhar M, Karamian M, Ghatee MA, Taylor WR, Ghohe HP, Rasooli SA. Distribution pattern of anthroponotic

cutaneous leishmaniasis caused by *Leishmania tropica* in Western Afghanistan during 2013-2014. Acta Tropica. 2017;**176**:22-28. DOI: 10.1016/j.actatropica.2017.07.028

[92] Ready PD. Leishmaniasis emergence in Europe. Euro Surveillance. 2010;**15**(10):29-39. DOI: 10.2807/ese.15.10.19505-en PMID: 20403308

[93] Fotakis EA, Giantsis IA, Avgerinou A, Kourtidis S, Agathaggelidou E, Kapoula C, et al. Identification of Leishmania species in naturally infected sand flies from refugee camps, Greece. Emerging Infectious Diseases. 2019;**25**:361-364. DOI: 10.3201/eid2502.181359 PMID: 30346269

[94] Gramiccia M, Gradoni L. The leishmaniases in southern Europe. In: Takken W, Knols BGJ, editors. Emerging Pests and Vector-Borne Diseases, Ecology and Control of Vector Borne Diseases. Vol. 1. Wageningen Academic Publishers; 2007. pp. 75-95

[95] Ready PD. Leishmaniasis emergence and climate change. In: de la Roque S, editor. Climate Change: The Impact on the Epidemiology and Control of Animal Diseases. Rev Sci Tech Off Int Epiz. Vol. 27, No. 2. 2008. pp. 399-412. DOI: 10.20506/rst.27.2.1803

[96] Van der Auwera G, Davidsson L, Buffet P, Ruf MT, Gramiccia M, Varani S, et al. LeishMan surveillance network. Surveillance of leishmaniasis cases from 15 European centres, 2014 to 2019: A retrospective analysis. Euro Surveillance. 2022;**27**(4):2002028. DOI: 10.2807/15607917.ES.2022.27.4.200202

[97] Reithinger R, Dujardin JC, Louzir H, Pirmez C, Alexander B, et al. Cutaneous leishmaniasis. The Lancet Infectious Diseases. 2007;**7**:581-596

[98] González C, Wang O, Strutz SE, González-Salazar C, Sánchez-Cordero V, et al. Climate change and risk of Leishmaniasis in North America: Predictions from ecological niche models of vector and reservoir species. PLoS Neglected Tropical Diseases. 2010;**4**(1):e585. DOI: 10.1371/journal.pntd.0000585

[99] Curtin JM, Aronson NE. Leishmaniasis in the United States: Emerging issues in a region of low Endemicity. Microorganisms. 2021;**9**:578. DOI: 10.3390/microorganisms9030578

[100] Arumugam S, Scorza BM, Petersen C. Visceral Leishmaniasis and the skin: Dermal parasite transmission to sand flies. Pathogens. 2022;**11**:610. DOI: 10.3390/pathogens11060610

[101] Germanó M, Salomón MC, Neira G, Lozano E, Mackern-Oberti JP, Cargnelutti DE. Leishmaniasis in Argentine Republic: Temporal and geographical distribution from 2013 to 2017. Asian Pacific Journal of Tropical Medicine. 2019;**12**(7):300-305

[102] Salomon OD, Quintana MG, Rosa JR. Eco-epidemiologia de leishmaniasis cutanea en Argentina. Sociedad Iberoamericana de Informacion Científica (SIICSALUD) Salud (i) Ciencia. 2008;**16**(5):514-520

[103] Hammami-Ghorbel H, Bem AI, Badri T, Chelbi H, Fenniche S, et al. Mucosal leishmaniasis of the lip: An emerging clinical form in Tunisia. Journal of the European Academy of Dermatology and Venereology. 2015;**29**:1212-1215

[104] Miguel DC, Guarnier DC. Canine and human Leishmaniasis: Disease progression to Brazilian urbanized areas. International Journal of Tropical Diseases. 2019;**2**:023

[105] Joao GG, Luz J, Victor LD, Amanda GC, Paulo AP, Juliana HC, et al. Human visceral leishmaniasis in Central-Western Brazil: Spatial patterns and its correlation with socioeconomic aspects, environmental indices and canine infection. Acta Tropica. 2021;**221**:105965

[106] Simão JC, Victória C, Fortaleza CMCB. Factors Affectng the spatial distribution of visceral Leishmaniasis in an urban area of recent emergence in inner Brazil. International Journal of Tropical Diseases. 2021;**4**:051. DOI: 10.23937/2643-461X/1710051

[107] Portella TP, Kraenkel RA. Spatial–temporal pattern of cutaneous leishmaniasis in Brazil. Infectious Diseases of Poverty. 2021;**10**:86. DOI: 10.1186/s40249-021-00872-x

[108] Filipe DT. Canine leishmaniosis in South America. Parasites & Vectors. 2009;**2**(1):S1. DOI: 10.1186/1756-3305-2-S1-S1

[109] Ramírez JD, Carolina H, Cielo ML, Martha SA, Carolina F, Camila G. Taxonomy, diversity, temporal and geographical distribution of cutaneous Leishmaniasis in Colombia: A retrospective study. Scientific Reports 2019;**6**:28266. DOI: 10.1038/srep28266

[110] Salgado-Almario J, Hernández CA, Ovalle CE. Geographical distribution of *Leishmania* species in Colombia, 1985-2017. Biomédica. 2016;**39**:278-290. DOI: 10.7705/biomedica.v39i3.4312

[111] Copa GN, María CA, Lorena VA, Alejandro JK, Silvana PC, Marisa J, et al. Transactions of the Royal Society of Tropical Medicine and Hygiene. 2019;**113**:91-100. DOI: 10.1093/trstmh/try113

[112] Anjalie A, Susiji WA. Comprehensive review of cutaneous Leishmaniasis in Sri Lanka and Identifcation of existing knowledge gaps. Acta Parasitologica. 2020;**65**:300-309. DOI: 10.2478/s11686-020-00174-6

[113] Adhikari RC, Shah M. Cutaneous leishmaniasis. Journal of Pathology of Nepal. 2017;7:1212-1217. DOI: 10.3126/jpn.v7i2.18031

[114] Ahammed A, Feroz S, Bhattacharyya B, Ahammed I, Selim S, Yusuf MA, et al. Living pattern and behavioural risk factors for kala azar: A case-control study in tertiary care hospital of Bangladesh. Bangladesh Journal of Infectious Diseases. 2016;**3**(2):29-35

[115] Abdullah AY, Ashraf D, Rakibul IS, Masudur R, Faruk H. Environmental factors associated with the distribution of visceral leishmaniasis in endemic areas of Bangladesh: Modelling the ecological niche. Tropical Medicine and Health. 2017;**45**:13. DOI: 10.1186/s41182-017-0054-9

[116] Alan JM, Wayne MM, Mary KKF, Ronald CN. Visceral Leishmaniasis. Standard Form 298 (Rev. 8-98) Prescribed by ANSI Std Z39-18

[117] Ghosh P, Roy P, Chaudhuri S, Das N. Epidemiology of post-kala-azar dermal leishmaniasis. Indian Journal of Dermatology. 2021;**6**(1):12. Available from: link.gale.com/apps/doc/A651095875/AONE?u=googlescholar&sid=googleScholar&xid=604c1edc. [Accessed: 24 July 2022]

[118] WHO. 2022. Available from: https://www.who.int/news-room/fact-sheets/detail/leishmaniasis

[119] Tangie LN, Desmond A, Aminde LN, Annabel MA, Halle PM. Cutaneous leishmaniasis in a severelyimmunocompromised HIV patient in Kumbo, northwest region of Cameroon: Case report. BMC Research Notes. 2017;**10**:425. DOI: 10.1186/s13104-017-2751-1

[120] Cloots K, Marino P, Burza S, Gill N, Boelaert M, Hasker E. Visceral Leishmaniasis-HIV coinfection as a predictor of increased Leishmania transmission at the village level in Bihar, India. Frontiers in Cellular and Infection Microbiology. 2021;**11**:604117. DOI: 10.3389/fcimb.2021.604117

[121] Lindoso JAL, Moreira CHV, Cunha MA, Queiroz IT. Visceral leishmaniasis and HIV coinfection: Current perspectives. Dovepress. HIV/AIDS - Research and Palliative Care. 2018;**10**:193-201. DOI: 10.2147/HIV.S143929

[122] Naveen SS, Doris PS. A review on major risk factors and current status of visceral Leishmaniasis in North India. American Journal of Entomology. 2019;**3**(1):6-14. DOI: 10.11648/j.aje.20190301.12

[123] Bhunia GS, Kumar V, Kumar AJ, Das P, Kesari S. The use of remote sensing in the identification of the eco-environmental factors associated with the risk of human visceral leishmaniasis (kala-azar) on the Gangetic plain, in North-Eastern India. Annals of Tropical Medicine and Parasitology. 2010;**104**:35-53

[124] Parmesan C. Ecological and evolutionary responses to recent climate change. Annual Review of Ecology, Evolution, and Systematics. 2006;**37**:637-669

[125] Oryan A, Akbari M. Worldwide risk factors in leishmaniasis. Asian Pacific Journal of Tropical Medicine. 2016;**9**(10):925-932

[126] Araujo AR, Portela NC, Feitosa APS, Silva OA, Ximenes RAA, Alves LC, et al. Risk factors associated with American cutaneous leishmaniasis in an endemic area of Brazil. Revista do Instituto de Medicina Tropical de São Paulo. 2016;**58**:86

[127] Terefe Y, Afera B, Bsrat A, Syoum Z. Distribution of human leishmaniasis (VL) and its associated risk factors, in Metemma, Ethiopia. Epidemiology Research International. **2015**:5. Article ID 630812. DOI: 10.1155/2015/630812

[128] Mandal PK, Wagle RR, Uranw S, Thakur AK. Risk factors for visceral leishmaniasis in selected high endemic areas of Morang, Nepal: A case control study. Journal of Kathmandu Medical College. 2020;**9**(4):188-196. DOI: 10.3126/jkmc.v9i4.36406

[129] Saha S, Ramachandran R, Hutin YJF, Mohan D, Gupte MD. Visceral leishmaniasis is preventable in a highly endemic village in West Bengal, India. Transactions of the Royal Society of Tropical Medicine and Hygiene. 2009;**103**:737-742

[130] Bamorovat M, Sharifi I, Aflatoonian MR, Sharifi H, Karamoozian A, Sharifi F, et al. Risk factors for anthroponotic cutaneous leishmaniasis in unresponsive and responsive patients in a major focus, southeast of Iran. PLoS One. 2018;**13**:e0192236. DOI: 10.1371/journal.pone.0192236

[131] Bamorovat M, Sharifi I, Aflatoonian MR, Sadeghi B, Shafiian A, Oliaee RT, et al. Host's immune response in unresponsive and responsive patients with anthroponotic cutaneous leishmaniasis treated by meglumine antimoniate: A case-control study of Th1 and Th2 pathways. International Immunopharmacology. 2019a;**69**:321-327. DOI: 10.1016/j.intimp.2019.02.008

[132] Kiptui EG, Sabella JK, Gladys JM. Risk factors of visceral leishmaniasis among residents of Baringo County, Kenya. International Journal of Community Medicine and Public Health. 2021;**8**(11):5251-5257. DOI: 10.18203/2394-6040.ijcmph20214257

[133] Murray HW, Berman JD, Davies CR, Saravia NG. Advances in Leishmaniasis. Lancet. 2005;**366**:1561-1577

[134] El Aasri A, Nizar S, Youssef El M, rutKhadija El K, Driss B. Sand flies (Diptera: Psychodidae) vectors of Leishmania: Notes on five species captured in Bni Oual locality, North-Western Morocco. Entomologie Faunistique – Faunistic Entomology. 2018;**71**:1-7

[135] Martínez DC, Ávila JL, Molano F. Sandfly (Diptera: Psychodidae: Phlebotominae) present in an endemic area of cutaneous leishmaniasis in West Boyacá, Colombia. Colombia Médica (Cali, Colombia). 2019;**50**(3):192-200. DOI: 1025100/cm.v50i3.3051

[136] Nawaz S, Irum S, Ahmed H, Shamas S, Roshan S, Raf MA, et al. Phlebotominae sandflies distribution and incidence of leishmaniasis from selected areas of Punjab Province, Pakistan. Punjab University Journal of Zoology. 2020;**35**(2):173-177. DOI: 10.17582/journal.pujz/2020.35.2.173.177

[137] Limoncu M, Emin İ, Cüneyt B, Seray T, Samiye D, Hakan K, et al. Entomological survey for the detection of sand Fly Fauna and Vector species in the cutaneous Leishmaniasis endemic area in East Mediterranean region of Turkey, Mersin Province. Journal of Medical Entomology. 2020;**57**(5):1510-1515. DOI: 10.1093/jme/tjaa089

[138] Pereira NCL, Érika MM, Fabiana OLS, Rosana SL, Adão JVP, Daniele MP, et al. Ecology of phlebotomine sand flies in a Brazilian area with recent leishmaniasis transmission (Itaúna, in Minas Gerais state). Journal of the Brazilian Society of Tropical Medicine. 2020;**53**:e20190538. DOI: 10.1590/0037-8682-2019-0538-2019

[139] Edna DD, Godwin KN, Seth OA, Alexander EY. Identification of sand flies (Diptera: Psychodidae) collected from cutaneous Leishmaniasis endemic focus in the Ho municipality, Ghana. Annals of Science. 2021;**10**(1):33-44. DOI: 10.21467/ias.10.1.33-44

[140] Al-Koleeby Z, Ahmed El A, Chafika F. Seasonal Changes of sandflies: A study of the Endemic Outbreak of Leishmania Major in Zagora, Southeast Morocco. E3S Web of Conferences. Vol. 234. 2021. p. 00104. DOI: 10.1051/e3sconf/202123400104

[141] Vadmal G, Caroline KG, Barbara AH, Bruno C, Adrian AC, Erin AM. Data-driven predictions of potential Leishmania vectors in Latin America. bioRxiv, The Preprint Server for Biology. 1-41. DOI: 10.1101/2022.08.22.504897

[142] Mahendra P, Iyasu E, Adugna G, Kushagra D, Pratibha D. Etiology, clinical Spectrum, epidemiology, diagnosis, public health significance and control of Leishmaniasis: A Compre-hensive review. Acta Scientific Microbiology. 2022;**5**(5):140-151. DOI: 10.31080/ASMI.2022.05.1066

[143] Umi A, Yudha N. New World Leishmania more Severe than Old World Leishmania: New World Species Cause a Spectrum of Disease Ranging from Mild Cutaneous Disease to Severe Mucosal Lesions. Research Gate; 2017. Available from: https://www.researchgate.net/publication/321422269

[144] Ondriska F, Bukovinova P, Votypka J, Nohynkova E, Boldis V. Imported new world cutaneous leishmaniasis in a traveller from Slovakia. Bratislavské Lekárske Listy. 2015;**116**(3):203-206. DOI: 10.4149/BLL_2015_040

Chapter 3

Epidemiology of Visceral Leishmaniasis in India

Rajan R. Patil and Prabir K. Chatterjee

Abstract

Kala-azar is a leading parasitic infection of great epidemic and mortality potential. More than 90% of Incident cases brought to notice of WHO in 2019 were mainly reported from 10 countries. Four endemic states in India namely Bihar, Jharkhand, Uttar Pradesh and Jharkhand have high disease burden of Kalazar. All 4 endemic states have to mandatorily notify cases to the National Vector Borne Disease Control Programme (NVBDCP) every month, even if there are zero cases. In recent years of Kala-azar cases, India have witnessed reduction of 97% largely due to the introduction of single-dose AmBisome –in India has been the game changer. There are three forms of leishmaniasis seen in India Kala-azar, they are Visceral Leishmaniasis, Post Kalazar Dermal Leishmaniasis (PKDL) and Cutaneous leishmaniasis. PKDL patients harbor the parasite and may be the source of new infection to the vector even 20 years later. Poverty enhances the risk for Kala-azar. Poor housing and domestic sanitary conditions are good breeding ground for sandfly which is the vector for Visceral Leishmaniasis, as well as resting sites and their ease of sandfly contact with humans. Kala-azar is a climate-sensitive disease as any change in temperature and humidity influences vector breeding.

Keywords: Kalazar, epidemiology, leishmaniasis, tribal population, India

1. Introduction

Kalazar is the widely used term for Visceral leishmaniasis (VL) in India. As per WHO Global Health Observatory, which is basically distribution maps of VL cases using the DHIS2 platform and has an interactive dashboard [1], VL is endemic in 78 countries. India, East Africa, and Brazil are the countries where most cases occur. Worldwide, about 50,000 to 90,000 new KA cases occur every year, however, only 25–45% are reported to WHO [2].

Kala-azar is a leading parasitic infection of great epidemic and mortality potential. More than 90% of incident cases that are brought to the notice of WHO in 2019 were mainly reported from 10 countries: India, Eiteria, Iraq, Somalia, Brazil, Ethiopia, Kenya, South Sudan, Sudan and Nepal. India's share of global burden of Kala-azar in 2020 has been 18%. Kala-azar remains a significant public health issue in India with 54 districts scattered in four states — Bihar (33 out of 38 districts), Jharkhand (4 out of 24 districts), Uttar Pradesh (6 out of 75 districts) and West Bengal (11 out of

23 districts). Stratified analysis showed that there are 633 blocks spread over four states which are endemic. Sporadic cases are also detected in states like Gujarat, Assam, Himachal Pradesh Haryana Tamil Nadu, Puducherry, Madhya Pradesh, Kerala, Sikkim, Uttaranchal and Jammu & Kashmir [3].

As per the instructions received from National Vector Borne Disease Control Programme (NVBDCP), All 4 states endemic to Kala-azar, namely Bihar, Jharkhand, West Bengal and Uttar Pradesh, have to mandatorily notify cases of the NVBDCP every month, even if there are zero cases. The spot map of Indian subcontinent (**Figure 1**) shows the distribution of various species of Leishmaniasis prevalent in India and its neighboring countries. Within India, Kalazar is clustering in eastern India as per data collected in last 2015 to 2019 [4].

Leishmania donovani is the pathogenic organism for Kalazar, and *Phlebotomus argentipes* is the main vector, particularly in India, commonly called as sand fly. Animal carriers are not known in India, though in a few countries, animal carriers [5] do exist in Europe e.g., foxes, and in China dogs are known to carry the infection. Disease spread of Kalazar is highly localized: over 3/4th of inferred infectors are seen to be at a mean distance <300 m to their respective secondary VL cases.

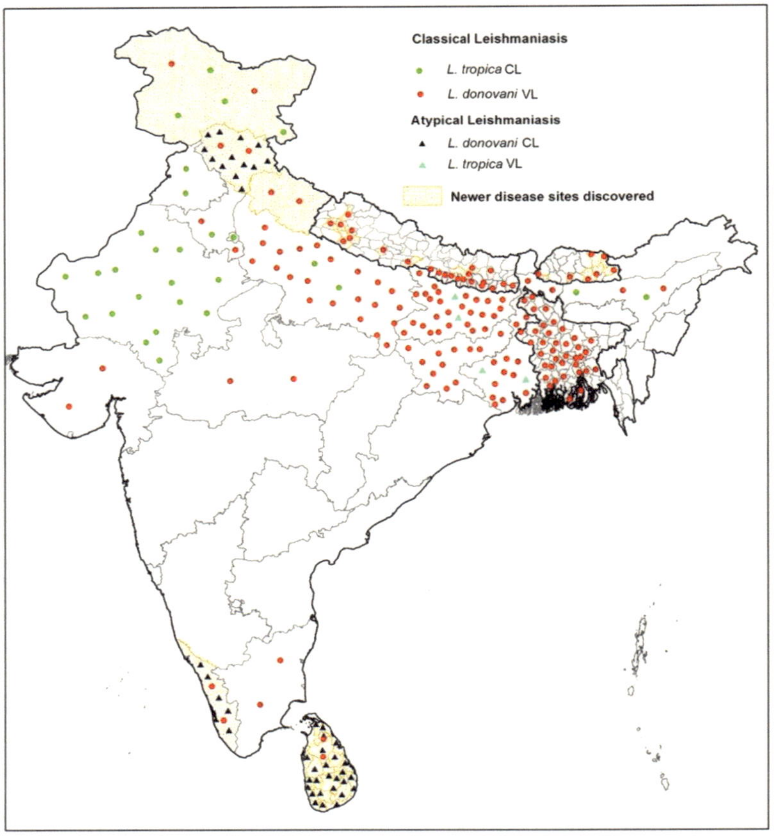

Figure 1.
Distribution of Kala azar in Indian subcontinent. Clustering of Kalazar cases in India and neighboring countries.

After many years (since 1992) of accelerated Kalazar elimination programme implementation, the cases of Kalazar cases in India have witnessed a reduction by 97%, while the reported deaths have fallen from 1419 in 1992 to 58 in 2018, and only 37 deaths were reported in 2020, formally [6].

Kalazar elimination as a public health problem is defined as achieving an annual incidence of less than 1 case per 10,000 population at the district level in Nepal and at the subdistrict level in Bangladesh and India.

National Kalaazar Elimination Programme NKEP, now integrated into National Vector Borne Disease Control Programme (NVBDCP) has set an ambitious goal of zero KA transmission by 2030. This goal will be attained through early diagnosis and prompt treatment, integrated vector management, operational research, effective disease surveillance, supervision, monitoring and evaluation and social/community mobilization and partnerships [7].

2. Kalaazar (visceral leishmaniasis)

Kalazar in its severe form is associated with high mortality in the absence of prompt diagnosis and treatment. Kalazar is prevalent in all the countries (**Figure 2**) in Indian subcontinent [4]. After Malaria, Kalazar comes second in ranking for causing maximum number of deaths by a parasitic disease. It is reported that 20,000

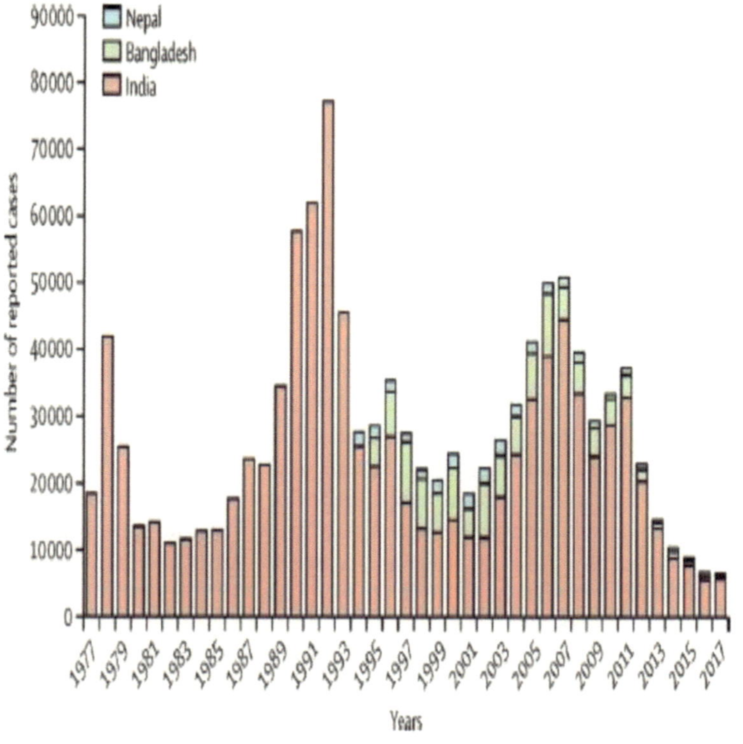

Figure 2.
Depiction of reported comparative reporting of Kala-azar cases in India, Nepal and Bangladesh from 1977 to 2017 [8].

Signs	Diagnosis
Loss of weight	Low hemoglobin
Wasting	Low white blood cell
Severe pallor	Low total count
Enlarged spleen	Low platelet count
	Bone marrow reveals Leishman-donovani (LD) bodies
	Molecular (PCR) Immunological (ELISA, rapid tests) assays. Other tests include rK39 strip test

Table 1.
Signs and diagnosis of Kalazar.

to 30,000 deaths occur worldwide due to Kalazar. The term Kalazar is derived from the word 'kala' which means black referring to black discoloration of skin in kalazar affected person. Since there is great similarity between the symptomatology of Kalazar and Malaria, the propensity to misdiagnose the disease is common. Misdiagnosis is perilous to the patient as the case fatality rate in Kalazar cases is 100% in the absence of proper treatment. In severe Kalazar, the death of the patient is not due to pathogen (L D bodies) but instead, due to opportunistic infection resulting from immune-compromised status due to Kalazar disease. The death by Kalazar is often due to secondary infections such as Dysentery, Pneumonia and Tuberculosis. which are highly prevalent in the communities where Kalazar is rampant [9, 10].

Symptoms of Kalazar are fever with chills. Generally, fever occurs twice a day. If not diagnosed and treated, the fever continues, though it is not so high after the initial days. Loss of appetite and severe weakness are other symptoms (**Table 1**).

3. Post-kala-azar dermal leishmaniasis

Post Kalazar dermal Leishmaniasis is a complication of Kala-azar, and is mainly seen in India, South-East Asia, and East Africa. It is manifested by discolored (hypopigmented) flat skin (macular) rash, combined with some slightly elevated (maculopapular) or elevated (nodular) rash. It is manifested in Kalazar patients who are cured of the VL. Manifestation of Post-kalazar dermal leishmaniasis (PKDL) usually occurs in the period of six months to one or more years after treatment and cure of Kala-azar, at the same time it may occur earlier or even along with Kala-azar. PKDL heals spontaneously in most cases in Africa but rarely in patients in India [11, 12]. PKDL patients harbor the parasite and may be the source of new infection in susceptible children even 20 years later.

In 6 districts of Bihar peak PKDL was in 2018 and in 2 districts it was in 2017. Recent papers suggest that the 2 districts achieved control around 2015 and elimination is likely between 2021 and 2023. In other districts it will not happen before 2022 given current spray rates (less than 50% in Kishanganj, between 60 and 70% in Gopalganj and Patna), endemicity (>2) and delays in treatment (60 days). PKDL was increasing in Jharkhand. There were 1800 reported cases of PKDL in India in 2020: 1034 were in Bihar, 502 in Jharkhand, 108 in UP and 156 in West Bengal. (NVBDCP, Ministry of Health, Govt of India.)

4. Cutaneous Leishmaniasis

Cutaneous Leishmaniasis is Less prevalent in India- sporadic cases are found in western India. Cutaneous manifestations are characterized by skin lesions mostly ulcers especially on exposed parts such as face, neck, and extremities. Skin lesions are the major reason for disability and stigma. As per the reported cases, 95% of cutaneous leishmaniasis worldwide comes from American continent, the Mediterranean region. There is gross under reporting of cutaneous leishmaniasis, it is understood that the actual number of cases may be in the range of 600,000 to 1,000,000, however only 200,000 cases are reported to WHO.

In the Indian region, Cutaneous Leishmaniasis is a highly localized public health problem in western India, restricted to one state of Rajasthan, especially in the Thar desert of Bikaner district. It is located at the center of Thar desert. The Bikaner city is surrounded by stone walls. Bikaner being a desert city expectedly the weather will be hot and dry with very little rainfall [13, 14].

Less common in India- mostly found in western India. CL cases were reported in 4 in Jammu, 2 in Punjab, 1 in Haryana, 1 in Delhi, 16 in Rajasthan, 2 in UP, 2 in Assam/Atypical CL 13 in Himachal, 2 in W Bengal, 5 in Kerala, 1 in Andhra, 2 in Maharashtra, 2 in Gujarat. (NVBDCP, Ministry of Health, Govt of India.)

5. Kala-azar-HIV co-infection

The chance of developing visceral Leishmaniasis clinical manifestation of disease accompanied by high relapse and death rates is very high in Kalazar-HIV coinfected cases due to Immuno- compromised status. Prompt treatment with antiviral treatment decreases the chance of disease development, keeping in check the relapses and enhancing the survival rates [15].

As of 2021, leishmania-HIV co-infection has been reported in 45 countries. Brazil, Ethiopia, and the state of Bihar in India have reported high Leishmania-HIV co-infection rates [16].

Due to the low immunity in Kalazar patients, many other co-infections, such as Tuberculosis and Dysentery, are commonly seen. Malaria and TB co-exist in Jharkhand.

Bleeding from Gums occurs due to low platelet count. Also seen is Noma (a Cancrum oris like growth) a known complication VL-HIV co-infection cases were 141 in 2018 and 91 in 2019 in Bihar. They were 7 in 2018 in Jharkhand and 2 in 2019. (NVBDCP, Ministry of Health, Govt of India.)

6. Treatment

Treatment of Kalazar is not simple, the drug administration should necessarily be done by trained health personnel since most of the anti-leishmanials are injectable drugs. All the Kalazar affected cases are required to comply with the drug regimen and take complete treatment [16].

Early diagnosis and prompt treatment of Kalazar result in decline in the prevalence of the Kalazar disease and hence reduction in the resultant disability and death. Epidemiologically, it helps in interrupting the chain of transmission subsequently reduction in disease burden in the community. Although anti-leishmanial drugs are costly, however highly subsidized procumbent by WHO has brought down the cost of medicine. Affordable medicines have helped improve access to essential medicine making in public health institutions. Good part is anti-leishmanials that are absolutely free to all the Kalazar patients [17, 18].

Liposomal Amphotericin-B (AmBisome) is the safest and effective injection. Other medicines include Sodium Stibonate (Antimony) Gluconate or SSG injections. Oral medicine Miltefosine, and Paramomycin (Injection)- which appear to be less effective.

6.1 Dosing

AmBisome is given as a Single dose in India and as 6/7 Doses in other countries (4 mg/kg/body wt.). SSG is given as 20 mg/kg body weight once a day for 30 days.

Miltefosine is given as 100 mg/day. Metclopramide – 5 mg/day is given along with Miltefosine.

Although single-dose AmBisome treatment is available in all endemic areas and satisfactory cure and relapse rates are reported (>95%). Significant delays between onset of symptoms and treatment were noted in Bihar and Jharkhand, due to the lack of diagnostic capacity at facility and case management.

7. Major risk factors

Socioeconomic conditions: Poverty enhances the risk for Kala-azar. Poor housing and domestic sanitary conditions are good breeding ground for sandfly which is the vector for Visceral Leishmaniasis, as well as resting sites and their ease of sandfly contact with humans. The mud plastered on a bamboo framework with cracks in the walls of houses is a typical breeding place. Lack of sunlight (due to there being no windows) is another feature as cool, dark spaces attract sandflies. Houses with crowding attract sandflies as these provide a good source of blood meals. Human behavior, such as sleeping outside or on the ground, may increase risk. Sandflies are zoophilic. So, keeping cattle inside the house and sleeping close by increases the chance of human beings getting Kalazar [19].

Malnutrition: Diets lacking in protein-energy, iron, vitamin A and zinc increase the risk that an infection will progress to a full-blown disease.

Population migration and movements of non-immune population in areas with active infection circulation are ideal conditions for epidemics of cutaneous as well as Kalazar.

8. Environmental changes

Kalazar is climate-sensitive as it affects the epidemiology in several ways:

- Vectors and reservoir get influenced by temperature changes, humidity, and rainfall by geospatial distribution which in turn is determinant of their survival and population size [20].

- Minor fluctuations in temperature could have a significant effect on the life cycle of *Leishmania* promastigotes in sandflies, facilitating transmission of the parasite in geographical places not previously endemic for the disease.

- Disasters like flood, drought and famine lead to massive human migration of population to newer areas that may or may not have Kalazar which may initiate fresh Kalazar transmission [21].

India has significantly improved Kalazar control programme. In endemic villages that have reported cases of Kala-azar over the past 3 years, 2 rounds of indoor residual spraying are being applied. Development of resistance in vectors has been the major challenge in efforts towards Kalazar elimination especially to pesticides such as DDT has led to the introduction of synthetic parathyroid for indoor residual spray in 2015 [22].

9. Public health intervention for Kalazar elimination

A series of public health measures have sustained the Indian Kala-azar elimination drive. These include:

- Capacity building of human resources and financial commitment;
- Case-based surveillance and micro-stratification in high-risk areas;
- For monitoring programme and initiate prompt action a new platform is introduced electronic health records to facilitate, KA Management Information System (KAMIS) [23];
- Introduction of new, single-dose liposomal amphotericin B treatment;
- In Kalazar-endemic areas, construction of concrete houses;
- Alphacypermethrin for Indoor Residual Spray (IRS) (replacement of DDT by 5%) since 2016;
- Introduction of superior hand compression pumps fitted with control flow valves for IRS (2015); Regular monitoring and evaluation of programmes by technical advisory groups, and collaboration with an effective network of partners.

The KA Management Information System (KAMIS) for NKEP which is essentially a web-based online platform has been in use since 2014. KAMIS holds electronic records of all KA and PKDL cases since 2013, along with data on drugs and diagnostics, IRS monitoring and vector surveillance. Data are entered in real-time at the implementation level by trained data operators.

To ensure prompt diagnosis and effective treatment, financial incentives are provided to community health workers, and wage losses have been compensated for KA and PKDL cases, with a good referral system with free of cost ambulance service. All Kalazar and PKDL patients are beneficiaries of the national health insurance scheme called Ayushman Bharat and avail absolutely free of cost health services including admission.

9.1 The reasons

Despite wonderful progress, last-mile challenges remain to eliminate Kala-azar. The reasons are varied ranging from factors such as difficult geographical terrain, indigenous populations, poor socio-economic conditions poor health-seeking behavior conditions, inadequate housing, and difficulties facing the implementation of the Kala-azar programme [24, 25].

With a progressive decline in cases, it is very important to maintain strong surveillance – this is now being integrated with those of other health surveillance programmes. Intensified governmental initiatives and political will has kept Kalazar elimination on high action alert.

10. Challenges in Kalazar elimination

There are several challenges that are encountered in attaining Kalazar elimination in India broadly they can be categorized as public health management, Biological factors & adverse impact pandemic on the operational aspect of Kalazar [26].

10.1 Challenges in surveillance and case management

Even though KA clusters are known, early diagnosis remains a challenge. KA symptoms are non-descriptive and nonspecific, ruling out other prevalent febrile diseases is the best way of finding KA-suspected cases. The first point of contact in endemic rural areas for Kalazar with fever is over-the-counter consultation with pharmacist or rural health practitioner, which is responsible for delay in treatment-seeking by 35–39 days at public health institutions. However, treatment-seeking delay at public health institutions ranged from 35 days in Bihar, however, it was lower in Jharkhand, 30 days. In lower endemic states the delay was much higher, 45 days in up and 50 days in West Bengal.

There are great hindrances in KA Surveillance when villages that have not previously reported KA begin to report KA cases from such villages, although they are located within known endemic blocks and districts. The number of such villages is declining each year, from 1001 in 2018 to 627 in 2019 and 87 in 2020.

10.2 KA elimination programme

Two significant challenges in KA–HIV coinfection and PKDL which are being obstacle to KA elimination. The percentage of KA cases with HIV has remained stable at 2–5%, while surveillance for PKDL has been fluctuating, with no known baseline of the disease burden.

Little is known about the outcomes of treatment of KA-HIV coinfected and PKDL cases. Certainly, the quality of diagnosis has improved significantly, however, identification of relapsed cases is being done in a few tertiary care health facilities [27].

10.3 Indoor residual spraying

According to national protocol, challenges in IRS as per guidelines, 2 rounds of IRS are conducted every year for sand fly control; however, the timing of IRS and the duration of spray rounds are affected by local contexts, IRS activity is affected by crop harvesting, elections, rains, migrations for work, all of which lead to delay in spraying activities or incomplete coverage [28].

As the spray pumps used in the programme are imported, obtaining spare parts (controlled flow valves, nozzle tips, lead gaskets, filter strainer) has been a challenge.

10.4 Impact of Covid-19 on Kalazar control programme

The COVID-19 pandemic has adversely impacted KA elimination programme since early 2020; despite the pandemic essential vector control and case management activities have been continued. The impact of COVID-19 on NKEP is yet to be assessed. With the shrinking geographical distribution of KA cases, having a strong surveillance system becomes paramount, in an effort to integrate KA case searches with those of other health programmes in a few KA-endemic states. Access to health care services especially for KA-HIV co-infected cases is challenge as they have to travel 160–240 km for treatment at a specialized tertiary hospital, Institutions of excellence are being set up to improve access health care to specialized care for complex KA cases and to build the trained health workers for case management [29].

Inter-sectoral Coordination among health and non-health departments and clarity in roles and responsibilities among all sectors, community mobilization will be key to sustaining the gains in the post-elimination phase [30].

11. Conclusions

Strengthening of Kalazar surveillance is key component of Kalazar elimination programme through strategic human resource planning in the health system. In the last leg of Kalazar elimination efforts have to be in mission mode. Both types of management styles are bottom-up as well as top to bottom. On paper, India has one of the finest national guidelines for the elimination of Kalazar, however, translating it to practical ground application with required resource mobilization would be primary pre-requisite. The use of quality epidemiological data is necessary to make an evidenced-based decision. Keeping with the spirit of national health policy there is a need for avoiding temptation of health systems to run Kalazar control programme as a vertical program, instead of integration with primary health care. Vector control against sand flies needs to take the holistic approach of integrated vector control methods.

Author details

Rajan R. Patil[1*] and Prabir K. Chatterjee[2]

1 KLE University, Belagavi, India

2 Amader Haspatal, Bankura, India

*Address all correspondence to: rajanpatil@yahoo.com

References

[1] The global health observatory [Internet]. Who.int. Available from: https://www.who.int/data/gho [Accessed: December 20, 2022]

[2] Leishmaniasis - India [Internet]. Who.int. Available from: https://www.who.int/india/health-topics/leishmaniasis [Accessed: December 20, 2022]

[3] Gill N, Pandey D, Roy N, Dhingra N, Jain S. Kala-azar in India -progress and challenges towards its elimination as a public health problem [Internet]. Available from: https://apps.who.int/iris/rest/bitstreams/1354276/retrieve

[4] Patil R. Independent Assessement of kala-azar Elimination Programm. New Delhi: Ministry of Health and Family Welfare, Government of India; 9-20 December 2019

[5] Medkour H, Davoust B, Dulieu F, Maurizi L, Lamour T, Marié J-L, et al. Potential animal reservoirs (dogs and bats) of human visceral leishmaniasis due to Leishmania infantum in French Guiana. PLoS Neglected Tropical Diseases. 2019;**13**(6):e0007456. DOI: 10.1371/journal.pntd.0007456

[6] Report of meeting of the regional technical advisory group (RTAG) on visceral leishmaniasis and the national visceral leishmaniasis programme managers of endemic member states [Internet]. Who.int. 2020. Available from: https://apps.who.int/iris/rest/bitstreams/1340195/retrieve [Accessed: December 20, 2022]

[7] Directorate National Vector borne disease control programme, Govt of India. Accelerated Plan for Kalazar Elimination. Directorate National Vector borne disease control programme, Govt of India; 2017

[8] Patil R. Surveillance Desk Review Paper for Independent Assessment of Kala Azar India. 2019

[9] Saini P, Kumar NP, Ajithlal PM, Joji A, Rajesh KR, Reena KJ, et al. Visceral Leishmaniasis caused by Leishmania donovani Zymodeme MON-37, Western Ghats. India. Emerg Infect Dis. 2020;**26**(8):1956-1958

[10] Burza S, Croft SL, Boelaert M. Leishmaniasis. The Lancet. 2018;**392**(10151):951-970

[11] WHO [Internet]. Who.int. Available from: https://apps.who.int/neglected_diseases/ntddata/leishmaniasis/leishmaniasis.html [Accessed: December 20, 2022]

[12] Leishmaniasis [Internet]. Who.int. Available from: https://www.who.int/health-topics/leishmaniasis [Accessed: December 20, 2022]

[13] Aara N, Khandelwal K, Bumb RA, Mehta RD, Ghiya BC, Jakhar R, et al. Clinco-epidemiologic study of cutaneous Leishmaniasis in Bikaner, Rajasthan, India. The American Journal of Tropical Medicine and Hygiene. 2013;**89**(1):111-115

[14] James WD, Berger TG, et al. Andrews' Diseases of the Skin: clinical Dermatology. Saunders Elsevier; 2006. p. 426

[15] Goswami RP, Goswami RP, Basu A, Ray Y, Rahman M, Tripathi SK. Protective efficacy of secondary prophylaxis against visceral Leishmaniasis in human immunodeficiency virus Coinfected patients over the past 10 years in eastern India. The American Journal of Tropical Medicine and Hygiene. 2017;**96**(2):285-291

[16] Bualert CS, Manomat J, et al. Risk factors of leishmania infection among hiv-infected patients in Trang province, southern Thailand: A study on three prevalent species. The American Journal of Tropical Medicine and Hygiene. 2020;**103**(4):1502-1509. DOI: 10.4269/ajtmh.20-0332

[17] Sundar S, Chakravarty J, Agarwal D, et al. Single-dose liposomal amphotericin B for visceral leishmaniasis in India. The New England Journal of Medicine. 2010;**362**(6):504-512

[18] WHO. Leishmaniais. Available from: https://www.who.int/india/health-topics/leishmaniasis#:~:text=The%20first%2Dline%20treatment%20for,and%20weight%20of%20the%20patient

[19] Singh RB et al. Breeding ecology of visceral leishmaniasis vector sandfly in Bihar state of India. Acta Tropica. 2008;**107**(2):117-120

[20] Bhowmick AR, Khanum H. Prevalence of visceral leishmaniasis, risk factors and associated disorders: Knowledge of inhabitants and professionals in Fulbaria, Mymensingh. Bangladesh Journal of Zoology. 2017;**45**:73-83

[21] Tidman R, Abela-Ridder B, De Castañeda RR. The impact of climate change on neglected tropical diseases: A systematic review. Transactions of the Royal Society of Tropical Medicine and Hygiene. 2021;**115**(2):147-168

[22] Valero NNH, Uriarte M. Environmental and socioeconomic risk factors associated with visceral and cutaneous leishmaniasis: A systematic review. Parasitology Research. 2020;**119**(2):365-384

[23] Who.int. Available from: https://www.who.int/news/item/29-07-2021-visceral-leishmaniasis-elimination-india-gears-up-to-overcome-last-mile-challenges [Accessed: December 20, 2022]

[24] KAMIS - Login page [Internet]. Gov.in. Available from: https://kamis.nvbdcp.gov.in/ [Accessed: December 20, 2022]

[25] Gedda MR, Bhawana Singh B, Kumar D, Abhishek Kumar Singh AK, Madhukar P, Upadhyay S, et al. Post kala-azar dermal leishmaniasis: A threat to elimination program. PLoS Neglected Tropical Diseases. 2020;**14**(7):e0008221

[26] Picado A, Ostyn B, Singh SP, Uranw S, Hasker E, Rijal S, et al. Risk factors for visceral Leishmaniasis and asymptomatic Leishmania donovani infection in India and Nepal. PLoS One. 2014;**9**(1):e87641

[27] Independent assessment of kala-azar elimination programme india [Internet]. Who.int. 2019. Available from: https://apps.who.int/iris/rest/bitstreams/1404094/retrieve [Accessed: December 20, 2022]

[28] Lindoso JAL, Moreira CHV, Cunha MA, Queiroz IT. Visceral leishmaniasis and HIV coinfection: Current perspectives. HIV AIDS (Auckl). 2018;**10**:193-201. DOI: 10.2147/HIV.S143929

[29] National Vector Borne Disease Control Progamme, Govt of India. Guidelines on vector control in kala-azar elimination. National Vector Borne Disease Control Progamme, Govt of India

[30] Paul A, Singh S. Visceral leishmaniasis in the COVID-19 pandemic era. Transactions of the Royal Society of Tropical Medicine and Hygiene. 2023;**117**(2):67-71. DOI: 10.1093/trstmh/trac100

Chapter 4

Leishmaniasis Epidemiology and Psychosocial Aspect

Ghulam Rahim Awab

Abstract

Leishmaniasis is a neglected tropical and the most important vector-borne disease caused by protozoan parasites of the genus *Leishmania*, transmitted by female sand-fly vector to the vertebrate host including human, highly correlated with poverty, malnutrition, climate and environmental factors such as crowded living conditions and poor sanitation that affect health, wellbeing, and livelihoods of millions of people around the world. Transmission is complex due to relationships between reservoir hosts, parasites, sand fly vectors, and socio-environmental risk factors. There are various clinical manifestations, ranging from spontaneously healing cutaneous lesions to potentially fatal visceral leishmaniasis caused by different *Leishmania* species. The psychological long-term manifestations leading to stigmatization, social exclusion, discrimination, and psychosocial impacts, advocating the importance of the One Health approach to combat these diseases effectively.

Keywords: leishmaniasis, neglected tropical diseases, vector-borne disease, epidemiology, sand fly, parasite

1. Introduction

Epidemiology of Leishmaniasis is the study of the characteristics of the parasite and sand-fly species, ecological characteristics of the transmission sites, current and past exposure of the human population to the parasite, human behavior, geographical distribution, and impacts on human populations [1]. Leishmaniasis is a multifactorial anthropo-zoonotic protozoal disease, responsible for various syndromes in humans, by way of anthroponotic and/or zoonotic transmission. The interactions among all of the agent, vector, climatic, environmental, and socioeconomic factors influence the risk of leishmaniasis [2]. Infection and transmission are initiated by inoculating of the causative agent (free promastigotes of *Leishmania*) by vector (female sand flies of the genus *Phlebotomus* and *Lutzomyia*) bite in vertebrate hosts [1]. The vertebrate hosts are attacked by female sand flies for blood meals necessary for eggs development cycle.

It is a global public health threat, a neglected grave tropical disease that portends possible poor prognosis and fatal consequences due to broad range of clinical manifestations observed worldwide with numerous endemic zones in various continents and cases are raising due to urbanization, deforestation [1, 3].

Leishmaniasis epidemiology has a major role in clarifying the etiology of particular *Leishmania* various species and variants as the cause of specific diseases, in improving our understanding of the overall characteristics of specific vectors, in determining factors affecting host susceptibility and immunity, in unraveling modes of transmission, in clarifying the interaction of parasite with environmental determinants of disease, in determining the safety, efficacy, and utility of preventive and curative weapons (personal protection, vaccines, drugs) especially in alerting and directing disease prevention and control actions [4]. Information on morbidity burden and biopsychosocial properties rates contributes directly to the establishment of priorities for prevention and control programs, whether this involves insecticide, drug and vaccine development and delivery, environmental and hygienic improvements, enhancement of socio-economic and nutritional status, personal or community behavior, agricultural and constructional processing enhancements, reservoir host and vector control, and international cooperation and communication.

2. Problem statement

Leishmaniasis is among the top three vector-borne diseases of parasite origin along with malaria and filariasis that impacts significantly on the health, wellbeing, and livelihoods of affected communities. It is also the third most common infectious disease in terms of morbidity, after tuberculosis and malaria with great diversity of parasite species, reservoirs, and vectors that play role in transmission [5]. The disease is considered as an endemic tropical and subtropical zoonotic health threat with outbreak and mortality potential occurs primarily in countries with (sub) tropical climates, disproportionally affecting individuals living with poverty, poor sanitation, malnutrition, restricted healthcare access, displacement, poor housing, weak immunity, and lack of financial resources. Existence in non-endemic areas linked to globalization, urbanization and environmental changes, migration, tourism and other international types of population movements [1]. Different species might cause different clinical features, and the severity differs from spontaneously healing lesions to life-threatening visceral disease that might be ended with more than 90% deaths if not promptly treated particularly in people living with HIV/AIDS [6]. The cutaneous leishmaniasis (CL) is the predominant clinical manifestation with skin lesions or ulcers, on the face, head, arms and legs and other exposed to sand-fly biting parts of the body [7, 8], leaving life-long scars and serious disability, disfiguration and stigma "horrific social discrimination" and cause significant maternal morbidity, and even fetal mortality during pregnancy [3, 9]. Although the disease is not lethal, the disfigurement and social stigmatization may cause or precipitate psychological disorders, restriction of social participation, reduction of health-related quality of life of affected individuals. Thus CL not only affects the physical well-being of the individual but also significantly alters their psychological, social, and economic well-being [10]. The overall leishmaniasis outcomes are determined by the inter-relationship between the individualities of the parasite, the bio-environmental aspects of the vector, and the immune response of host [1].

3. Problem magnitude

Globally, leishmaniasis is among the top ten neglected tropical health burden, endemically existing in people of every continent except Australia and Antarctica.

Currently, 98 countries are endemic for leishmaniasis in the tropics, subtropics, and Southern Europe. More than 350 million individuals are at risk, 12 to 15 million infected, and 1.5 to 2 million new cases occur annually with 20,000–30,000 deaths [11, 12]. The endemic regions include Mediterranean Basin, Africa, Central and South America and Asia. As per WHO 2012 report, predictably, 59 countries are endemic for visceral lieshmaniasis, although cases fall excessively on seven of these as 90% of these cases happened in INDIA, South Sudan, Brazil, Ethiopia, Kenya, Sudan, Somalia and among 95% of CL cases occurred in the Americas, Middle East and Central Asia, approximately 70% reported in Syria, Brazil, Algeria, Colombia, Afghanistan, Iran, Ethiopia, Sudan, and Costa Rica [13].

Presently, over 85% of new CL cases occurred in 10 countries, namely Afghanistan, Algeria, Brazil, Colombia, Iraq, Libya, Pakistan, Peru, the Syrian Arab Republic, and Tunisia. Likewise, 10 countries: Brazil, China, Ethiopia, Eritrea, India, Kenya, Somalia, South Sudan, Sudan, and Yemen were accountable for more than 90% of new VL cases reported to WHO [11].

4. Epidemiological determinants

It is known that different species of *Leishmania* can cause in many mammalian hosts various clinical manifestations, and the severity varies from spontaneously healing lesions to life-threatening visceral disease. The outcome is determined by the inter-relationship between the characteristics of the parasite, the biology and habitat of the vector, and the behavior and immune response of host (**Table 1**) [1].

4.1 Agent

Leishmaniasis human disease is caused by more than 20 pathogenic species of *Leishmania* parasites transmitted from host to host by a tiny insect. *Leishmania* parasites are categorized into either old world (The Eastern Hemisphere) or new world (The Western Hemisphere) species, most notably the old world species: *L. major, L. tropica, L. aethiopica*, and the new world species: *L. amazonesis, L. braziliensis, L. mexicana, L. panamensis, and L. guyanensis* [15]. Old world species can be found in Asia, the Middle East, the Mediterranean Basin, and Africa, whereas new world species are found in the Americas [16].

4.2 Vector

Approximately, in over all 600 known species of sand flies, 10% act as disease vectors. And only 30 are important from public health point. Particular species of *Leishmania* are transmitted by particular adult female sand fly of *phlebotomine* belonging to either *Phlebotomus* spp. (Old World) or *Lutzomyia* spp. (New World) to either incriminated or suspected reservoir hosts [17]. The sand flies generally are active at night and inoculate the parasite when they bite humans ("from dusk to dawn") [18] and are less active in the hottest day-time, but still bite if they are bothered. Sand flies do not make noise, they are small, and their bites might not be noticed the parasite are spread. Adult sand flies often inhabit rock cracks, hollows, and rodent holes, and rest in cool, dark, and humid corners of animal shelters or human dwellings and their ability to acquire, maintain, and

Agent	Vector	Reservoir/Host	Geographical location
Old World			
Labrus donovani	*P. Argentipes, P. orientalis, P. martini, P. alexandri*	Humans (anthroponosis)	North-East India, Bangladesh, Nepal, Sudan, Kenya, Horn of Africa, China
L. infantum	*P. perniciosus, P. perfiliewi, P. chinensis, etc.*	Dogs, cats, foxes, jackals	Mediterranean, Sudan, West Africa, Middle East, China, Central Asia
L. major	*P. papatasi, Polypauropus duboscqi*	Gerbils (e.g. Rhombomys, Meriones), Rodents, Human anthroponosis	Semideserts in North Africa and Middle East, North India, Pakistan, Central Asia, Sub-Saharan Savannah, Sudan
L. tropica	*Paralongicollum sergenti*,	Humans, dogs	Mediterranean Basin, Central Asia, North Africa, Middle East, Iran, Afghanistan
L. aethiopica	*P. longipes, P. pedifer*	Hyraxes (Procavia, Heterohyrax), Rodents	Highlands of Kenya, Ethiopia, Uganda
New World			
L. infantum	*Lu. longipalpis, Lu. evansi*	Dogs, foxes, opossums (Didelphis)	Most of Central and South America, especially Brazil
L. mexicana	*Lu. Olmeca*	Forest rodents (especially Ototylomys)	Central and Northern South America
L. amazonensis	*Lu. Flaviscutellata*	Forest rodents (especially Proechimys, Oryzomys)	Tropical forests of South America
L. brasiliensis	*Lu. wellcomei, Lu. whitmani, etc*	Rodents, opossums, dogs, and equines	Tropical forests & cultivated land throughout South & Central America
L. guyanensis	*Lu. Umbratilis*	Sloths (Choleopus), arboreal anteaters (Tamandua)	Northern South America
L. panamensis	*Lu. trapidoi, etc.*	Sloths (Choleopus)	Central America, Ecuador, Colombia
L. peruviana	*Lu. verrucarum, Lu. Peruensis*	Dogs, rodents, opossums	West Andes of Peru

Abbreviations P = Phlebotomus, Lu., Lutzomyia.

Table 1.
Disease types and transmission cycles of leishmaniasis worldwide [14].

transmit the parasites depends on maturity, density, biting habit parasite-vector-host interactions as well as on the ecological and epidemiological features of the infection [19]. Many sand fly species are opportunistic and feed on easy to access animals [20].

4.3 Reservoir hosts (host preference)

Humans, domestic animals, and wild animals are known to act as reservoir hosts such as up to 70 animal species, have been found as natural reservoir hosts of Leishmania parasites.

4.3.1 Human

Despite the human genetic susceptibility, hereditary factors, population ethnicity, genome-wide association, and immunological responses correlated to disease epidemiology [21, 22] the following demographic factors are of consideration.

Age: People of all ages are at risk for leishmaniasis if they live or travel to leishmaniasis endemic areas. Indigenous prevalence increases with age up to 15 years, after which the prevalence stabilizes or decreases, presumably reflecting the progressive buildup of immune protective status [23, 24]. Children under 10 years old displayed the highest proportion (>50%) of new infection [25] potentially due to lack of protective immunity or due to them collecting water or playing in gorges close to sand-fly habitat.

Gender: There is no real gender difference in leishmaniasis, but for decades, male patients have comprised the majority of reported cases, perhaps because of referral bias, more risk-taking behaviors by men or possibly, men are less likely to strictly adhere to recommended preventive measures and thus increase their risk of contracting the disease [26, 27]. The sex-related differences cannot be explained solely in terms of environmental exposure, socio-cultural determinant or healthcare access. Furthermore, transcriptomic evidence is revealing that biological sex is a variable impacting physiology, immune response, drug metabolism, and consequently, the progression of disease [28].

Poverty: Poverty increases the risk for leishmaniasis due to remote and rural inhabitance, low-income, malnutrition, poor waste management (such as open sewerage), poor housing and domestic sanitary conditions which increase sand-fly breeding and resting sites, as well as their access to humans. Sand flies are attracted to crowded housing as these provide a good source of blood-meals. Human behavior, such as sleeping outside or on the ground, may increase risk. Malnutrition and protein deficit, as well as a lack of iron, zinc, and vitamin A in the diet is also a risk factor for complications and for the disease progressing to kala-azar [29].

Population mobility: Migration, tourism, shipment to an area of endemic illness, and occupational exposure of non-immune people into areas with existing transmission cycles increase the risk of infection and epidemics leishmaniasis are often associated with massive migration, internal dislocation, and the population movement [30, 31].

4.3.2 Domestic animals

Dogs are the known reservoirs for the parasites of all types of leishmaniasis and natural host for *L. infantum, L. chagasi, L. tropica, and L. peruviana* as being infected by them, prevalently in Mediterranean, Asia, and Latin America [32]. Cats are considered secondary reservoirs of the infection in endemic areas as Leishmaniasis has been reported sporadically in several parts of the world [33, 34].

Cattle may possibly increase transmission burden by remaining as untreated reservoir for the parasite, their closeness to human owners, and because of increased availability of blood meals for the sand fly [35].

4.3.3 Wild animals

The detection of Leishmania infection in rodents, suggests that wild animals are also contributing to maintaining the life cycle and transmission of Leishmaniasis. Several rodents, namely, Mus musculus (domestic mouse), Microtus socialis, Rattusrattus (black rat), Cercomys cunicularius (wild rat), Mesocricetus auratus (Syrian hamsters), and marsupial e.g. Didelphis albiventris (opossum) in America, Africa, and Asia lead to spread of leishmaniasis [36, 37]. Other potential reservoirs are squirrels, wild canids, reptiles, and bats [38–40].

4.4 Ecology/environment

Leishmaniasis is climate-sensitive such as the pathogens, vectors, and hosts involved in the transmission cycle are environmentally sensitive [41] and impacted by climate change and human modification of ecosystems, therefore, changes in ecological settings (range from rain forests to deserts), rainfall, atmospheric temperature, and humidity resulting in fluctuations in sand fly numbers [42] due to variation in the developmental time and metabolism of sand flies and Leishmania development cycle within vector [43]. Humidity and moisture from rainfall or in the soil dealing with survival, development, and activity of sand flies. The matter of hygiene, (which is generally neglected in rural areas), presence of waste or stables in the vicinity of houses, use of traditional building materials for house construction, are all environmental factors that favor the development of the vector and reservoirs, therefore impact the incidence and distribution of leishmaniasis in different regions of the world. Environmental changes such as conversion of natural forest to other land uses, habitat destruction, altered landscape composition, urbanization, human incursion into forested areas, tourism into natural areas, the creation of roadways, energy networks, new farmlands, water management, and poorly planned urban development play an important role in Leishmaniasis epidemiology [44].

5. Transmission of pathogens

Understanding the transmission cycles is very important in effective prevention of leishmaniasis. There is no direct transmission from person to person and the indirect (vector-born) transmission of Leishmania species that cause VL or CL can be zoonotic or anthroponotic [45] depends on the characteristics of the parasite and sand-fly species, characteristics of the transmission sites, current and past exposure of people to the parasite, and human behavior. The zoonotic transmission from canine to humans, is found in the Mediterranean region and many other drier regions of Latin America. Leishmania species reported from dogs and occasionally from cats include *L. mexicana, Labrus donovani*, and *L. braziliensis*. Cats are at risk of infection especially in areas where these parasites are endemic [19]. In many endemic areas, infected people are not necessarily maintaining the transmission cycle, but infected animals along with sand flies, maintain the cycle. Only for anthroponotic transmission, infected people are needed to maintain the cycle (human-sand fly-human), parasite transmission being mostly nocturnal and typically seasonal [46].

The incubation period in human is quite variable, generally 1 to 4 months; range is 10 days to 3 years. Visceral leishmaniasis is 3–8 months (range 10 days to 34 months) and of the Cutaneous leishmaniasis is 2 weeks to several months (rarely up to 3 years).

6. Psycho-social impact

Leishmaniasis due to disfigurement and social stigmatization causes or predisposes severe psychological disorders and negatively affects the quality of life varied from minor domestic restrictions to severe physical and emotional isolation that results in poor mental health [47]. The adolescents with lesions are often faced with difficulty in social relations such as finding job, friends, spouse or deemed unsuitable for marriage or social communal life. While the children disfigured by lesions or scars, due to painful treatment, because of exclusion from play with other children are also psychologically affected and emotionally damaged [48].

The social exclusion, discrimination, and psychosocial impacts negatively affect socioeconomic opportunities which showed that depression and anxiety symptoms were higher in patients. Body satisfaction was also impaired in the groups with active CL and healed scars. Culturally, older people were more accustomed to the CL scar, but younger generations had less acceptance of any lifelong stigma and disfigurement on the face [49].

Understanding the psychosocial consequences, economic cost, health importance and the interaction of pathogenic protozoan, sand flies vectors, environmental impact is encouraging prevention, control, and surveillance of Leishmaniasis under a holistic view and integrated approach of human and veterinary medicine, environmental science, and wildlife conservation under the "One Health" approach.

Author details

Ghulam Rahim Awab[1,2]

1 Medical Faculty, Nangarhar University, Jalalabad, Afghanistan

2 Oxford Tropical Diseases Research Unit (MORU Tropical Network), Mahidol, Bangkok, Thailand

*Address all correspondence to: awabgr@yahoo.com

References

[1] Kammona O, Tsanaktsidou E. Nanotechnology-aided diagnosis, treatment and prevention of leishmaniasis. International Journal of Pharmaceutics. 2021;**605**:120761

[2] Valero NNH, Uriarte M. Environmental and socioeconomic risk factors associated with visceral and cutaneous leishmaniasis: A systematic review. Parasitology Research. 2020;**119**(2):365-384

[3] Speich B, Croll D, Fuerst T, Utzinger J, Keiser J. Effect of sanitation and water treatment on intestinal protozoa infection: A systematic review and meta-analysis. The Lancet Infectious Diseases. 2016;**16**(1):87-99

[4] Organization WH. Control of Neglected Tropical Diseases: World Health Organization. WHO; 2021

[5] Burza S, Croft SL, Boelaert M. Leishmaniasis. Lancet. 2018;**392**(10151):951-970

[6] Silva B da, Barreira J, M. Beatriz P.P. Oliveira. Natural phytochemicals and probiotics as bioactive ingredients for functional foods: Extraction, biochemistry and protected-delivery technologies. Trends in Food Science & Technology. 2016;**50**:144-158

[7] Arroub H, Hamdi S, Ajaoud M, Habbari K, Lemrani M. Epidemiologic study and molecular detection of Leishmania and sand fly species responsible of cutaneous leishmaniasis in Foum Jamaa (Azilal, atlas of Morocco). Acta Tropica. 2013;**127**(1):1-5

[8] El Hamouchi A, Ajaoud M, Arroub H, Charrel R, Lemrani M. Genetic diversity of Leishmania tropica in Morocco: Does the dominance of one haplotype signify its fitness in both predominantly anthropophilic Phlebotomus sergenti and human beings? Transboundary and Emerging Diseases. 2019;**66**(1):373-380

[9] Pires M, Wright B, Kaye PM, da Conceição V, Churchill RC. The impact of leishmaniasis on mental health and psychosocial well-being: A systematic review. PLoS One. 2019;**14**(10):e0223313

[10] Rafferty J. Curing the stigma of leprosy. Leprosy Review. 2005;**76**(2):119-126

[11] Organization WH. Leishmaniasis. Fact Sheet N 375. Updated January 2014. Geneva, Switzerland: World Heal Organ; 2014

[12] Torres-Guerrero E, Quintanilla-Cedillo MR, Ruiz-Esmenjaud J, Arenas R. Leishmaniasis: A review. F1000Research. 2017;**6**:750

[13] GHO. By category. Leishmaniasis [Internet]. Available from: https://apps.who.int/gho/data/node.main.NTDLEISH?lang=en

[14] Pigott DM, Bhatt S, Golding N, Duda KA, Battle KE, Brady OJ, et al. Global distribution maps of the leishmaniases. eLife. 2014;**3**:e02851

[15] Kevric I, Cappel MA, Keeling JH. New world and old world Leishmania infections: A practical review. Dermatologic Clinics. 2015;**33**(3):579-593

[16] Mans DR, Kent AD, Hu R, Schallig H. Epidemiological, biological and clinical aspects of leishmaniasis with special emphasis on Busi Yasi in Suriname. Journal of Clinical & Experimental Dermatology Research. 2017;**8**(02):388

[17] Maroli M, Feliciangeli MD, Bichaud L, Charrel RN, Gradoni L. Phlebotomine sandflies and the spreading of leishmaniases and other diseases of public health concern. Medical and Veterinary Entomology. 2013;**27**(2):123-147

[18] Murray HW, Berman JD, Davies CR, Saravia NG. Advances in leishmaniasis. Lancet. 2005;**366**(9496):1561-1577

[19] Ready PD. Biology of phlebotomine sand flies as vectors of disease agents. Annual Review of Entomology. 2013;**58**:227-250

[20] Abbasi I, Cunio R, Warburg A. Identification of blood meals imbibed by phlebotomine sand flies using cytochrome b PCR and reverse line blotting. Vector-Borne Zoonotic Diseases. 2009;**9**(1):79-86

[21] Castellucci LC, Almeida L, Cherlin S, Fakiola M, Carvalho E, Figueiredo AB, et al. A Genome-wide association study highlights a regulatory role for IFNG-AS1 contributing to cutaneous leishmaniasis in Brazil. bioRxiv. 2020

[22] Chakravarty J, Hasker E, Kansal S, Singh OP, Malaviya P, Singh AK, et al. Determinants for progression from asymptomatic infection to symptomatic visceral leishmaniasis: A cohort study. PLoS Neglected Tropical Diseases. 2019;**13**(3):e0007216

[23] Douba MD, Abbas O, Wali A, Nassany J, Aouf A, Tibbi MS, et al. Chronic cutaneous leishmaniasis, a great mimicker with various clinical presentations: 12 years experience from Aleppo. Journal of the European Academy of Dermatology and Venereology. 2012;**26**(10):1224-1229

[24] Reithinger R, Dujardin J-C, Louzir H, Pirmez C, Alexander B, Brooker S. Cutaneous leishmaniasis. The Lancet Infectious Diseases. 2007;7(9):581-596

[25] van Henten S, Adriaensen W, Fikre H, Akuffo H, Diro E, Hailu A, et al. Cutaneous leishmaniasis due to Leishmania aethiopica. EClinicalMedicine. 2018;**6**:69-81

[26] Malaria Consortium C. Leishmaniasis control in eastern Africa: Past and present efforts and future needs. Situation and Gap Analysis. 2010:86

[27] Lezama-Davila CM, Oghumu S, Satoskar AR, Isaac-Marquez AP. Sex-associated susceptibility in humans with chiclero's ulcer: Resistance in females is associated with increased serum-levels of GM-CSF. Scandinavian Journal of Immunology. 2007;**65**(2):210-211

[28] Lockard RD, Wilson ME, Rodríguez NE. Sex-related differences in immune response and symptomatic manifestations to infection with Leishmania species. Journal of Immunology Research 2019;2019.

[29] Organization WH. Leishmaniasis: Fact Sheet N 375, Updated January 2014. WHO; 2014

[30] Fakhar M, Ghohe HP, Rasooli SA, Karamian M, Mohib AS, Hezarjaribi HZ, et al. Genetic diversity of Leishmania tropica strains isolated from clinical forms of cutaneous leishmaniasis in rural districts of Herat province, Western Afghanistan, based on ITS1-rDNA. Infection, Genetics and Evolution. 2016;**41**:120-127

[31] Charyyeva A, Çetinkaya Ü, Özkan B, Şahin S, Yaprak N, Şahin I, et al. Genetic diversity of Leishmania tropica: Unexpectedly complex distribution pattern. Acta Tropica. 2021;**218**:105888

[32] Gramiccia M. Recent advances in leishmaniosis in pet animals: Epidemiology, diagnostics and anti-vectorial prophylaxis. Veterinary Parasitology. 2011;**181**(1):23-30

[33] Soares CSA, Duarte SC, Sousa SR. What do we know about feline leishmaniosis? Journal of Feline Medicine and Surgery. 2016;**18**(6):435-442

[34] Longoni SS, López-Cespedes A, Sánchez-Moreno M, Bolio-Gonzalez ME, Sauri-Arceo CH, Rodríguez-Vivas RI, et al. Detection of different Leishmania spp. and Trypanosoma cruzi antibodies in cats from the Yucatan peninsula (Mexico) using an iron superoxide dismutase excreted as antigen. Comparative Immunology, Microbiology and Infectious Diseases. 2012;**35**(5):469-476

[35] Hirve S, Boelaert M, Matlashewski G, Mondal D, Arana B, Kroeger A, et al. Transmission dynamics of visceral leishmaniasis in the Indian subcontinent–a systematic literature review. PLoS Neglected Tropical Diseases. 2016;**10**(8):e0004896

[36] Pourmohammadi B, Motazedian MH, Kalantari M. Rodent infection with Leishmania in a new focus of human cutaneous leishmaniasis, in northern Iran. Annals of Tropical Medicine and Parasitology. 2008;**102**(2):127-133

[37] Lainson R. The American leishmaniases: Some observations on their ecology and epidemiology. Transactions of the Royal Society of Tropical Medicine and Hygiene. 1983;**77**(5):569-596

[38] Kassahun A, Sadlova J, Benda P, Kostalova T, Warburg A, Hailu A, et al. Natural infection of bats with Leishmania in Ethiopia. Acta Tropica. 2015;**150**:166-170

[39] Abebe A, Evans DA, Gemetchu T. The isolation of Leishmania aethiopica from the ground squirrel Xerus rutilus. Transactions of the Royal Society of Tropical Medicine and Hygiene. 1990;**84**(5):691

[40] Gebresilassie A, Yared S, Aklilu E. Sandfly fauna and ecological analysis of Phlebotomus orientalis and Phlebotomus martini in the lowland foci of visceral leishmaniasis in Somali regional state, Southeast Ethiopia. Asian Pacific Journal of Tropical Medicine. 2020;**13**(1):31

[41] Purse BV, Masante D, Golding N, Pigott D, Day JC, Ibañez-Bernal S, et al. How will climate change pathways and mitigation options alter incidence of vector-borne diseases? A framework for leishmaniasis in south and Meso-America. PLoS One. 2017;**12**(10):e0183583

[42] Elmazini S, Ejghal R, Bekhti K, Lemrani M. The sporadic cutaneous leishmaniasis due to Leishmania infantum in Morocco: A presumably trend towards endemicity. Acta Tropica. 2021;**227**:106288

[43] Kholoud K, Denis S, Lahouari B, El Hidan MA, Souad B. Management of leishmaniases in the era of climate change in Morocco. International Journal of Environmental Research and Public Health. 2018;**15**(7):1542

[44] Semenza JC, Suk JE. Vector-borne diseases and climate change: A European perspective. FEMS Microbiology Letters. 2018;**365**(2):fnx244

[45] Assefa A. Leishmaniasis in Ethiopia: A systematic review and meta-analysis of prevalence in animals and humans. Heliyon. 2018;**4**(8):e00723

[46] Alten B, Maia C, Afonso MO, Campino L, Jiménez M, González E, et al. Seasonal dynamics of phlebotomine sand fly species proven vectors of Mediterranean leishmaniasis caused by Leishmania infantum. PLoS Neglected Tropical Diseases. 2016;**10**(2):e0004458

[47] Hofstraat K, van Brakel WH. Social stigma towards neglected tropical diseases: A systematic review. International Health. 2016;**8**(Suppl. 1): i53-i70. DOI: 10.1093/inthealth/ihv071

[48] Pal B, Murti K, Siddiqui NA, Das P, Lal CS, Babu R, et al. Assessment of quality of life in patients with post kalaazar dermal leishmaniasis. Health and Quality of Life Outcomes. 2017;**15**(1):1-7

[49] Chahed MK, Bellali H, Ben Jemaa S, Bellaj T. Psychological and psychosocial consequences of zoonotic cutaneous leishmaniasis among women in Tunisia: Preliminary findings from an exploratory study. PLoS Neglected Tropical Diseases. 2016;**10**(10):e0005090

Chapter 5

Cutaneous Leishmaniasis

Azhar Rafique, Sayydah Surrya Sani, Salma Sultana, Tayyaba Sultana, Asma Ashraf and Muhammad Shahid Mahmood

Abstract

Cutaneous leishmaniasis (CL) is a widespread parasitic infection caused by the Leishmania, which is carried by female sandflies. The symptoms include basic ulcer to lethal systemic disease *i.e.*, formation of widely dispersed skin lesions of diverse types. Almost 350 million individuals are at danger and the disease is endemic in more than 98 countries. There are globally 12 million cases, with 2–2.5 million new cases annually. Cutaneous leishmaniasis is considered as critically neglected disease by WHO. Earlier it was difficult to identify the infecting parasite, but modern DNA techniques make it quite simple to identify the Leishmania species, allowing quick treatment decisions. The quick identification of Leishmania is made possible using the PCR method. There is currently no vaccination to prevent leishmaniasis, and pharmacological treatment is frequently ineffectual. There is a need for broad and well-conducted investigations to help its control. Amphotericin B, pentamidine isethionate, paromomycin, and antifungals are some of the drugs recommended for treatment. By organising direct, in-person training, which is a crucial step in improving attitudes and preventative actions toward CL and its control in endemic areas, it is necessary to underline the significance and necessity of teaching this at-risk population.

Keywords: cutaneous leishmaniasis, protozoa, diagnosis, PCR, identification

1. Introduction

In 190 developing nations, leishmaniasis, an infectious illness brought on by protozoa of the genus Leishmania, continues to be a significant public health issue [1]. Leishmania parasite infection can result in one of three main clinical forms, depending on the species that caused the infection. *Leishmania infantum* and *L. chagasi* have been found to be identical by genotyping [2] *Leishmania chagasi* is considerate a sub-population of *L. infantum* that arose from imported European strains [3–5], therefore they should be regarded as synonyms and are presents in both NW and OW. The first is localised cutaneous leishmaniasis (CL), which can cause a single or numerous skin ulcers as well as satellite lesions or nodular lymphangitis. CL with mucosal involvement is the third kind, systemic visceral leishmaniasis (VL), which affects internal organs such the liver, spleen, and bone marrow and is lethal if untreated. CL without mucosal involvement (CL) is the second kind [6].

Males are more prone than females to get this sickness [7]. The leishmaniasis disease is spread by phlebotomine flies, of the genera Phlebotomus in the ancient world and Lutzomyia in the new world. The vector of this disease belongs to the order Diptera, class insecta, Family: Psychodidae [8]. The sandfly has a 3 mm length and is known for its "hopping" flight. They feature fragile, long legs, dagger-shaped mouthparts, huge, dark eyes, long antennae, and downward-facing mouthparts [9].

By being bitten by female sandflies, 20 different Leishmania species cause his illness. The disease is spread by 30 different species of sandflies. Humans and non-wild or domesticated animals serve as their reservoir hosts. Female sandflies consume reservoir hosts and contract the disease [10]. Use of contaminated syringes from sick people is how this parasite is spread. *Leishmania chagasi, Leishmania infantum,* and *Leishmania donovani* are the common species [11]. Due to the shift of individuals toward urban areas during the past 20 years, there has been a potential increase in leishmaniasis cases [12]. Leishmaniasis spread among people living in non-endemic areas through travel.

Just seven nations—Algeria, Afghanistan, Brazil, Iran, Peru, Syria and Saudi Arabia, —represent 90% of all CL cases. CL is the most common clinical type of leishmaniasis worldwide [13]. The Old World species of Leishmania parasites, such as *L. infantum*, *L. tropica* and *L. major* (common in, the Middle East, the horn of Africa, the Mediterranean basin and the Indian subcontinent), and the New World species, such as *L. chagasi*, *L. mexicana*, *L. amazonensis, L. naiffi*, (endemic in Middle and South America). Self-limiting ulcers are frequently caused by Old World species, but American tegumentary leishmaniasis, which also causes MCL and disseminated cutaneous leishmaniasis (DCL), is typically caused by New World species [14].

Leishmaniasis patients have a wide immunological spectrum, from those with a robust T cell response, as shown by delayed-type hypersensitivity (DTH) and high levels of interferon γ (IFNγ), to those who lack a DTH response but still have high antibody levels [11]. Others with a robust DTH have few parasites in their lesions because Leishmania spp. are killed by IFNγ -activated macrophages instead of being neutralised by antibodies, but people with merely a humoral response are unable to manage the parasite count [15, 16].

In addition to varying clinical symptoms based on species, Leishmania species also differ in their sensitivity to treatment options [17]. Because of this, determining the species is crucial to how leishmaniasis will manifest clinically. The identification of the Leishmania parasite that causes the disease used to be difficult, in disparity to many other infectious diseases. With the development of new DNA techniques, Leishmania parasites can now be recognised quite easily. With the motto "Small bite, big threat," the World Health Organisation (WHO) highlighted the serious and growing risk of vector-borne diseases, particularly leishmaniasis, on World Health Day 2014 [18]. In order to improve vector management, diagnosis, and the treatment toolset to prevent additional incidence and morbidity, leishmaniasis is considered a category 1 emergent and uncontrolled sickness and requires more intensive study. The major subjects of this review are the diagnosis, management, prevention, and strategies for the management and control of CL caused by both Old World and New World species.

2. Sandfly and Leshmania life cycle

In the blood meal of several hematophagus arthropods, the eaten amastigotes (protist cell, non-motile) in the infected host change into promastigotes (external

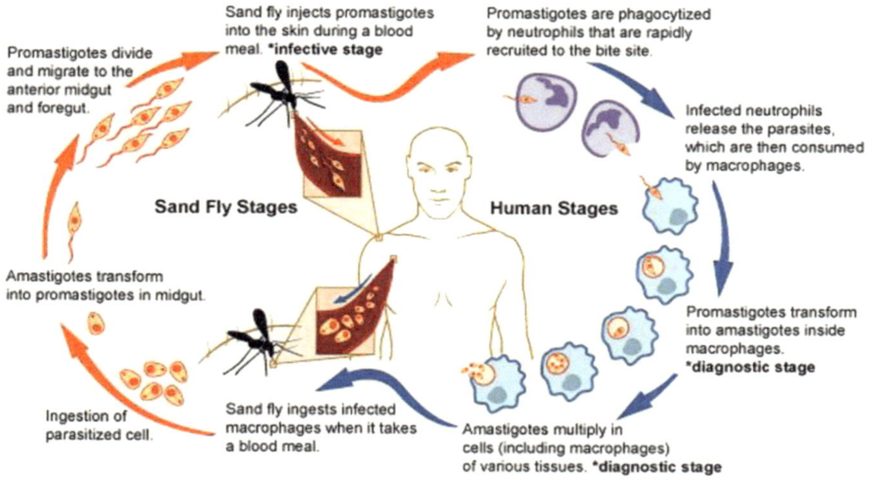

Figure 1.
Leishmaniasis' life cycle (http://www.niaid.nih.gov/topics/leishmaniasis/pages/lifecycle.aspx).

flagellum). When a host is inappropriate, parasites are discharged with the excrement [19]. When the blood meal is digested, trypanosomatids such as Leshmania species and others bind to the parasites on the midgut epithelium. The parasite is then kept in the gut and begins a stage differentiation [20]. Promastigotes then proceed to Sandfly's foregut, which is covered in cuticles, of the sandfly. In the foregut some attach and some remain free for next transmission by bite [21] into the vertebrate host.

Promastigotes reproduce by binary fusion in the sandfly's digestive system. After seven days, promastigotes undergo metacyclogenesis and become contagious (metacyclic promastigotes). The sandfly will puncture the host's skin while feeding in order to inject saliva and metacyclic promastigotes into the host (**Figure 1**).

Metamorphose into the amastigote form within the host macrophage. They multiply inside the phagolysosome by binary fission until the cell ruptures and spreads the infection to additional phagocytic cells. This keeps the cycle going.

3. Early defence against Leishmania

The several species of Leishmania that cause cutaneous leishmaniasis each have unique characteristics. However, the parasites do share a life cycle in which a sand fly transmits a promastigote, a flagellated form of the parasite, to mammalian hosts such as humans, dogs, and rodents [22]. The promastigotes enter various phagocytic cells after being introduced into the skin by a sand fly's bite. Promastigotes change into an amastigote, a circular, non-flagellated replicative form, inside the phagolysosome of macrophages. When sand flies consume amastigotes while feeding on a host, the amastigotes then change into promastigotes and reproduce within the sand fly, the life cycle is complete. However, during a natural infection, other elements found in the sand fly saliva are delivered in the skin that affects early immune responses. Most experimental infections involve injecting promastigotes into the skin with a needle [23]. The biological significance of studies looking at the early response to infection without taking into account the conditions present during a natural infection, such as the inoculation site, the number of parasites, and the components present during the sand fly bite, must therefore be carefully interpreted [24, 25].

While neutrophils, dendritic cells (DCs), and monocytes that are recruited to the infection site also have a chance of contracting the parasite and play critical and distinctive roles in the establishment of the immune response to infection.

4. Causes of leishmaniasis

The most prevalent type of leishmaniasis that affects people is cutaneous leishmaniasis. It is a cutaneous condition spread by the bite of a phlebotomine sand fly and brought on by a single-celled parasite. Leishmania species that can lead to cutaneous leishmaniasis number around thirty.

5. Signs and symptoms (diagnosis)

5.1 Cutaneous leishmaniasis

Skin scrapings samples—often collected from the advantage of lesions—that have been microscopically examined are typically used to make the diagnosis. Although quick and inexpensive, this has a low sensitivity, especially in persistent lesions [26]. Although, bacteria and fungi present in the biopsy samples may contaminate the cultures of the lesions. Additionally, the needs for growth vary among species. Isoenzyme electrophoresis can be used to identify specific Leishmania species, although the procedure is time-consuming, expensive, and requires extensive parasite cultivation. Although PCR is preferable for a direct investigation of clinical material, monoclonal antibodies can also be employed to identify species in cultured strains. High specificity and sensitivity PCR is quick. Leishmania detection and genetic characterisation are also possible at the same time [27]. The PCR sensitivity for one investigation on American cutaneous leishmaniasis was 100% [28]. Due to the absence of considerable antibody generation in cutaneous leishmaniasis, antibody detection is not very sensitive.

Additionally, in cases of American cutaneous leishmaniasis, reports of cross-reactivity between leishmanial antigens and antibodies made by other kinetoplastids, such as *Trypanosoma cruzi*, have been made [29]. The Montenegro (leishmanin) skin test, which looks for a particular type of delayed-type hypersensitivity on the skin, is another method for diagnosing cutaneous leishmaniasis that is currently available. *L mexicana* antigen is injected intradermally while being watched for a native reaction [30]. This test's limitations include the difficulty to distinguish between an infection that is currently present and one that has already occurred, as well as instances of false positive results in cases of other skin conditions [31].

5.2 Mucocutaneous Leishmaniasis

Mucocutaneous leishmaniasis develops after the commencement of cutaneous leishmeniasis and is defined by the destruction of the pharyngeal, oral, and nasal canals. The occurrence of this disease is also significantly influenced by genetic factors. Although nasal inflammation and stuffiness are the main early signs of mucocutaneous leishmaniasis, the septum may slowly perforate and ulcerate. The larynx, mouth, throat, soft palate, and face are all affected by the lesion [32]. Bacterial infections may not harm the bones, but untreated illness can cause diarrhoea, pneumonia, and tuberculosis [33]. Suffocation, lung diseases, and starvation are other factors that might lead to death (due to closure of laryngeal aperture) [34].

5.3 Post kala-azar dermal leishmaniasis

A recurrence of kala-azar known as post-kala-azar dermal leishmaniasis (PKDL) may show up on a person's skin months or even up to 20 years after being partially treated, left untreated, or even in those who were thought to have had adequate treatment [35, 36]. In up to 60% of instances that are being treated in Sudan, they can be proven. They appear as facial redness or hypopigmented skin lesions (such macules, papules, and nodules). Even though PKDL can be caused by any organism that causes kala-azar, it is frequently linked to *Leishmania donovani*, which causes various disease patterns in Sudan and India. Nodules from the African type frequently ulcerate as they advance, whereas nodules from the Indian variant frequently expand with time and create plaques but seldom do so. Involvement of the nerves is more common in the African variation than in the Indian subcontinent [37]. Histology shows a variety of chronic inflammatory cells; macrophage or epitheloid granuloma may be present [38]. The variation in parasite concentration throughout investigations may be due to the less sensitive diagnostic techniques utilised in previous entries.

6. Clinical features

6.1 Scratches

An infection brought on by a sandfly bite can either remain asymptomatic or show up as an expanding, ulcerating papule after an incubation period of 1 to 12 weeks. An average scratch or lesions resembles an ulcer that is not painful and has a raised, indurated edge and a necrotic base that is frequently coated in an adherent crust of dried exudate. Most patients have 1 or 2 lesions, typically on exposed areas that range in diameter from 0.5 to 3 cm [39]. However, there is a great deal of variation: whereas some lesions develop sporotrichoid nodular lymphangitis, others do not ulcerate. Pain can be brought on by typical secondary bacterial infections. Atrophic scars are left behind after the majority of lesions heal over several months to years. *L. tropica* lesions typically take longer to heal—about 10 months—than those caused by *L. major* or *L. mexicana*, and *L. brazilinesis* lesions typically last considerably longer [40, 41]. Partial resistance to reinfection results from natural resolution [42].

In a study of 475 cases of *L. major* in Saudi Arabia, the parasite was found in 50–80% of smears (depending on the researcher and the technique employed), 70% of skin biopsies, and only 50% of cultures. In 10–20% of cases, even after combining all three procedures, the parasite remained undetectable. In New World illness, the issue is frequently worse, especially in lesions older than six months [43].

6.2 Alternative methods

Alternative methods for obtaining tissue for diagnosis include needle aspirates and slit skin smears. A needle aspirate is obtained using a 2 ml leur lock syringe, a 20 gauge needle, and 0.3 ml 0.9% saline. The needle is inserted through healthy skin and 0.1 ml is injected into the margin of the lesion. The tiny tissue fragments are aspirated after being severed from the needle track's edge while the needle is being rotated back and forth and suctioned. To create smears and inoculate cultures, use the aspirate. Slit-skin smears are made by pinching the edge of the lesion between the thumb and fingers, making a 1 mm deep slit with a scalpel, and then scraping the cut edge [44].

Because antibodies frequently go undetected or are present in low, serology is ineffective for treating cutaneous diseases [45]. Similar to the tuberculin test, the Leishmanin skin test identifies cell-mediated immunity; it turns positive after the scratches start to crust and stays that way for ever. It is unable to differentiate between current and previous infections [46].

7. Morphology

A papule that resembles an insect bite marks the beginning of the lesion. The papule persists and gradually grows in size rather than regressing. The nodular, noduloulcerative, and ulcerative forms are three common morphological presentations in the following stage. The latter features a flat foundation with marginal rolls that are either small or significant. If it is a secondary infection, the base is made up of granulation tissue and may be covered in pus.

Satellite papules, which are tiny (2–4 mm) papules along the perimeter of the lesion, are another crucial aspect of morphology. Skin crease orientation is visible in the lesions. Multiple lesions frequently cluster together; the sand fly bites the same spot repeatedly, delivering parasites with each bite.

Special clinical features of *L. major* infection are: [47].

1. Diffuse thickening without ulceration (erysipeloid form).
2. Leishmanial cheilitis.
3. Chiclero's ulcer; this is an ulcer on the pinna of the ear.
4. Mycetoma-like lesions.
5. Leishmanial dactylitis.
6. Cutaneous leishmaniasis of the nose.
7. Sporotrichoid cutaneous leishmaniasis: the infection spreads along the lymphatics as in sporotrichosis.

Leishmaniasis recidivans, also known as lupoid leishmaniasis, is a unique clinical form of *L. tropica* infection; in this situation, the sore appears to heal but recurs along the edge of the lesion, a process that may last for many years and may be disfiguring. Like cutaneous tuberculosis, it (lupus vulgaris). Typically, there are few parasites, and the LST is quite positive.

8. Infections in animals

8.1 Clinical signs

8.1.1 Dog

Wild dogs and domestic dogs are the main reservoirs of zoonotic visceral leishmaniasis caused by *L. infantum* in the Mediterranean area, Middle East, Asian countries, and Latin America. Dogs serving as the main reservoir of visceral leishmaniasis have made research into the immune response and searching for Leishmania antigens linked to protective cellular immunity in canine visceral leishmaniasis more attractive. Recent studies have shed fresh light on the genetic underpinnings, pathophysiology, immunology, and epidemiology of canine leishmaniasis. These new discoveries have improved understanding of the condition and aided in the creation of novel diagnostic techniques and infection-control strategies, such as dog collars impregnated with insecticide, new medications, and second-generation vaccines [48, 49].

Other infected dogs may remain asymptomatic or display one or more minor illness, which is known as oligosymptomatic infection [50]. Some infected dogs may experience symptoms that result in death. The distinctive histological feature in the skin, liver, and spleen is a granulomatous inflammatory response connected to Leishmania amastigotes within macrophages [51].

The clinical indications of canine visceral leishmaniasis were used to categorise a group of mixed-breed dogs with spontaneous Leishmania infections as symptomatic or asymptomatic [52].

Domestic dogs in Latin America have been observed to naturally contract *L. braziliensis, L. peruviana, L. panamensis, Leptodactylus colombiensis, and L. mexicana* [48]. There is currently no conclusive proof that dogs serve as reservoir hosts for the domestic spread of CL [53, 54]. The majority of research focuses on determining the prevalence of CL in dogs, but little is known about the parasitic and immunological aspects of the infection.

8.1.2 Cats

Leishmaniasis can occur in cats, but most infected cats are believed to remain asymptomatic. Most frequently, lesions of the skin or mucosa are described, either with or without visceral symptoms. Visceral symptoms, however, might appear without cutaneous involvement. Skin lesions can be found anywhere on a cat, although most frequently appear on the lips, nose, ears, eyelids, and paws. The most frequent lesions encountered are localised papules, nodules and chronic crusted or ulcerated lesions; regional lymphadenopathy may also be present. Rare reports of alopecia, scales, and hemorrhagic pustules or nodules have been made. Initial lesions are frequently single, although they can sometimes be numerous and occasionally spread.

There have been reports of oral, nasal, and in certain cases other mucous membranes (such the anal mucosa) being involved. Some cats can have ocular symptoms, including unilateral or bilateral uveitis, conjunctivitis, and blepharitis (which can develop into panophthalmitis). Fever, hepatomegaly, jaundice, vomiting, diarrhoea, lymphadenopathy, dyspnea, nasal discharge, anaemia, and leukopenia are just a few of the visceral abnormalities and symptoms that have been observed in cats [55].

8.1.3 Equidae

Skin lesions can occasionally appear on horses, mules, and donkeys, especially on the head, neck, legs, and inguinal or axillary regions. Solitary or numerous papules or nodules, which frequently have ulcers, are the most typical lesions. Additionally, widespread skin illness has been documented. Although visceral leishmaniasis in horses has not been documented, parasites and *L. braziliensis* nucleic acids have been found in the blood and bone marrow of other animals in South America [55].

8.2 Other domestic animals

Rarely have medical instances in cattle or other small ruminants been reported. Only skin lesions, occasionally accompanied by lymphadenopathy, were noted in goat, sheep and cattle. In Germany, a pregnant cow with *Lechytia martiniquensis* infection had several ulcerative or plaque-like skin lesions on various body parts. After giving birth, it fully recovered. The only clinical symptom in experimentally infected sheep was a fever. Pigs with an experimental infection exhibited no symptoms.

Rarely, *L. enriettii* skin lesions, frequently on the ear, have been found in naturally infected guinea pigs. The earliest lesions in experimentally infected animals start out as redness and swelling but quickly progress into sizable, ulcerated lumps that resemble tumours. While some investigations discovered secondary lesions at other locations, including as the skin, lip, and genitalia, others claimed the lesions did not spread. Additionally, parasites were found in various internal organs. Spontaneous healing has been documented in certain studies but not others. Hamsters formed non-ulcerated nodules that went away on their own after contracting *L. enriettii* infection in an experiment.

8.3 Captive wild species and wild animals

The few known cases of leishmaniasis in wild or captive canids have resembled leishmaniasis in canines. Some nonhuman primates have been found to have visceral involvement with nonspecific symptoms (such pale mucous membranes and weight loss). A lion displayed clinical symptoms of colitis, including bloody diarrhoea, epistaxis, weight loss, and footpad sores.

L. martiniquensis-infected captive Australian marsupials have experienced skin lesions that include elevated, crusty or ulcerative pale nodules as well as localised to converging patches of thickened skin. Some rats infected with the *L. mexicana* complex in the wild have been reported to have skin lesions. These lesions are described as ulcers or swollen bumps with thinning hair. The tail base was said to be where they appeared most frequently, but they might also appear on the ears or toes. Numerous species have been known to have subclinical infections [55].

9. Treatment

Luckily, there are some rules to follow. A wait-and-see strategy for spontaneous cure may be acceptable because most lesions heal quickly without therapy, especially for patients who live in endemic areas because spontaneous healing is linked to the development of protective immunity. It is recommended to actively treat lesions that are multiple or persistent, have associated lymphangitis, are on cosmetically or functionally significant places like the hands or face, have several lesions. Patients with early, non-inflamed lesions should have local therapy; patients with many lesions or more complex lesions should receive systemic therapy.

9.1 Sodium stibogluconate and Meglumine antimoniate

Patients with cutaneous leishmaniasis can receive an intralesional infiltration of 1–5 ml (100 mg/ml) sodium stibogluconate (SSG) on alternate days for three days once a month, and in the majority of instances, this causes full healing by the end of the second month [56]. For numerous and larger lesions, a higher dose and more than three regimens were required. In addition to intralesional SSG in several lesions, Sharma et al. employed intramuscular SSG (800 mg/day) [56]. Meglumine antimoniate, a substitute medication that is the top drug of choice in Ecuador [57]. In Nepal, sodium antimony gluconate (20 mg/Kg/day) is administered intramuscularly to PKDL patients for duration of 30 to 72 days [58].

Drug resistance is a problem, and cases of visceral leishmaniasis with SSG resistance were reported in Nepal [59, 60]. So far, this issue was not reported from Nepal in cutaneous leishmaniasis. It is unknown that what is the most effective treatment for American utaneous and mucocutaneous leishmaniasis (ACML)? Since the 1940s, pentavalent antimonial medications have been utilised, such as sodium stibogluconate (SSG) and meglumine antimonate (Glucantime, MA), but they are costly, poisonous, and uncomfortable [61]. It is advised to identify the precise species of Leishmania before beginning treatment because medications that are effective for one species of Leishmania may not be effective for another. Sadly, leishmaniasis is an orphan disease in affluent countries, and nearly all of the available treatments are poisonous and have serious adverse effects [61].

Compared to parenteral therapy, this results in higher local concentrations and fewer systemic side effects. 5–10 1–5 ml antimony infiltrations are performed two–three times per week. Make sure the injection is in the lesion and not in the tissues beneath the skin. It might be really painful. There is growing evidence that it is useful in treating *L. tropica* infections, as well as CL brought on by *L. major* [62].

10. Physical methods

Patients of all ages have received treatment for cutaneous leishmaniasis using a variety of physical techniques, such as surgical excision, cauterization, cryotherapy, and the use of local heat.

10.1 Cryotherapy

Cryotherapy involves applying liquid nitrogen repeatedly to a lesion up to 2 mm outside the lesion margin using a cotton-tipped applicator or a cotton swab with moderate pressure. Each application's freezing time is 15–20 seconds. The process is carried out two or three times with brief breaks, taking between 30 and 120 seconds in total. The whitening of the skin at 2–3 mm outside the lesion's edges indicates proper treatment [63–66]. The typical post-freeze pattern includes blistering of the lesion for two to three days, crusting, mild oedema and the development of an eschar [63–66].

11. Treatment in animals

Animals rarely receive topical treatments, but radio-frequency-induced heat therapy was effective in treating two dogs with several localised mucocutaneous lesions on the snout. In certain animals, such as some cats and some horses, cutaneous lesions did not recur following surgical resection; but, in other instances, surgical resection alone was ineffective.

11.1 Vector control

Theoretically, infection might be controlled by preventing transmission by the sandfly vector. Depending on the species, some sandflies are endophagic and eat indoors, while others are exophagic and eat outdoors. Examples of tactics include deterrents, particularly pyrethroids, and insecticides like DDT. The majority of the times, sandflies are still quite sensitive to insecticides, despite reports of DDT resistance. Spraying has been shown effective at the local level, but it is unclear what effect blanket spraying would have on the sandfly population, and these programmes are challenging to maintain. Bednets offer defence against species that consume living things [67] with pyrethroid-impregnated nets providing extra protection dropping biting rates by up to 64–100% [68]. The success of bednets as a long-term control method depends on routine, replacing damaged nets, re-impregnation and dissemination to rural areas. The ability of bednets to stop the spread of malaria to children has raised incentive to improve this method of malaria control, including research into bednets treated with long-lasting insecticides [68].

11.2 Animal reservoirs

Leishmaniasis is a zoonotic disease that is widespread throughout the world with significant reservoirs of infection in sylvatic and domestic animals. The dog population has been targeted throughout the Mediterranean basin and Brazil. The use of dog collars impregnated with delta methrin has proven to be the most successful tactic, offering dogs up to 86% protection during high transmission seasons [69]. Dog killing has also been used, and modelling from a Brazilian study suggests that in areas with low endemicity, both dog killing and dog collars should have a greater proportional impact. However, as transmission rates rise, the relative benefit of dog collars decreases [70]. However, the effectiveness of vector and animal reservoir control programmes has been diminished due to decreased pesticide use and rising prevalence among urban populations. Given the wide range of epidemiological situations, the multiplicity of factors that affect disease transmission, and the continuous understanding regarding the biology of the parasite, its vector, and its reservoir hosts, disease management has thus proven to be extremely difficult to achieve. To avoid these problems, a vaccine for humans or dogs is a preferable option.

11.3 Vaccines

A plausible basis for vaccine development can be found in the discovery that strong immunity against re-infection accompanies spontaneous or medication-induced recovery from CL or VL. This finding gave rise to the long-standing technique of "leishmanization," or the use of live parasites retrieved from skin lesions to cause lesions in desired body regions in order to prevent sickness on re-infection.

Such a custom has been practised for at least 2000 years. Between 1982 and 1986, Iran had about 1.2 million patients receive this live vaccination [71]. 93% of people who had skin lesions after receiving the vaccination had a positive leishman-delayed hypersensitivity skin test, which is a good field predictor of population immunity. Skin lesions occurred in almost 50% of vaccination recipients. Additionally, a considerable decrease in disease incidence was seen, going from 14% in the group that had not gotten the immunisation to 2.5% in the group that had. Although it has not always produced cross-species protection, the use of heterologous organisms with lower pathogenicity as vaccines against a more virulent species is justified by the greater extent of immunological cross-reactivity between species at the humoral and cellular levels. Leishmanization is no longer used due to the risk of localised sickness and HIV-related spread, as well as the impossibility of delivering new cultures of a live vaccination in the field. A different approach with attenuated organisms enables exposure to a considerably wider spectrum of antigens than is possible with more complex subunit vaccines, resulting in the establishment of an immune response that is most similar to that of a natural infection. However, even using naturally pathogenic organisms in such a way to treat human or experimental murine leishmaniasis [72], exposed organisms [73] or genetically manipulated organisms [74], there has been little achievement. Similar to this, dead vaccinations have low immunogenicity and effectiveness even when supplemented with adjuvants, such as bacille Calmette-Guérin BCG or alum [75]. Interestingly, BCG alone caused a positive leishmanin skin test in some people, most likely because Mycobacteria and Leishmania have antigenic cross-reactivity. While the effectiveness of the single-dose *L. major* vaccination combined with BCG in dogs was around 70% [76]. It has showed promise as a possible vaccination policy against a natural *L. infantum* infection to administer to dogs naturally excreted secretory antigens extracted from culture supernatant of *L. infantum* promastigotes [77]. TSA, LeIF, and LmSTI1 are three recombinant leishmanial antigens that have shown immunogenicity in canine [78]. Recent developments in the creation of a vaccination that prevents transmission are hopeful given the significance of dogs as virus reservoirs. 92–97% protection against zoonotic visceral leishmaniasis was produced by the *Labrus donovani* fucose-mannose ligand (FML) antigen in combination with saponin (FML vaccination and Leishmune), with considerable protection shown out to 12 months [79]. Investigating the use of specific compounds as human vaccinations is an alternative strategy. Animal models for a gp63 peptide vaccination were successful [80]. The failure to adequately elicit cellular immunity, a necessary component for the management of intracellular infections, has, nevertheless, contributed to the generally poor effectiveness of these vaccines in humans.

11.4 DNA vaccines

The creation of DNA vaccines has provided an innovative solution to this issue. According to Wolff et al., intramuscular injection of plasmid DNA expressing a variety of reporter genes could cause muscle cells to produce proteins [81]. This study provides a solid basis for the concept that pure recombinant nucleic acids can be delivered in vivo to regulate protein production. Later studies found that DNA vaccines could defend mice from CL [82]. The ability to retain information over the long term is one of the main determinants of a vaccine's effectiveness. DNA encoding leishmanial antigen LACK is better to leishmanial protein and IL-12 protein immunisation for maintaining antigen-specific Th1 responses that can restrict *L. major* infection [83]. Antigen persistence and IL-12 activation by CpG motifs are two factors that explain why DNA vaccination is more effective than protein and adjuvant. In an experiment with leishmaniasis, a study finding that anti-IL-10R antibody resulted to sterile cure but loss of immunity against re-infection showed that parasite persistence was essential for the preservation of immunity [84]. However, DNA vaccines can nonetheless induce long-lasting immunity in the absence of noticeable antigen. This is done either by allowing undetectable antigen to remain, potentially in follicular DCs, or by inducing antigen-independent immune responses [85]. According to studies, memory T cells are diverse, with one fraction (central memory T cells) migrating through lymph nodes and another migrating to tissues and producing effector cytokines [86]. Recent research shows that even in the absence of parasites, central memory T cells can mediate long-term memory [87]. Finding the pathways that cause central memory T cells to be induced would therefore be a significant problem for DNA vaccines. The potential for improving DNA vaccine immunogenicity has been investigated in a number of ways. By acting as TLR ligands, novel adjuvants including CpG motifs and monophosphoryl lipid A release IL-12 and encourage a Th1 response. Heterologous primeboost immunisation has been used as another method to improve human responses to vaccination. Various combinations have been studied, but priming with DNA and boosting with MVA have received the most attention [88]. This work demonstrates that IL-10 plays a crucial role in the immune response by showing that the efficacy of the vaccine against these antigens in the BALB/c model was dictated by IL-10 from regulatory T cells. The most frequently sampled gene among expressed sequence tags from cDNA libraries of *L. major* was TryP, according to early investigations [89], in vulnerable BALB/c mice as a reproducibly protective antigen against infection. LACK, *L. major* stress-inducible protein 1, Leishmania elongation factor, and HASPB1, a stage-specific hydrophilic acylated surface protein in mice, are other antigens that are effective in mice and non-human primates. In order to immunise against the saliva-containing *L. major* challenge, the use of salivary antigens in plasmid DNA has also been examined [90]. The 8500 discovered genes now that *L. major* Friedlin's genome sequence is complete serve as a source of potential vaccine candidates [91]. There is no presumed requirement that the target antigen be a surface molecule because the parasite is inside, which significantly expands the pool of potential vaccine candidates. One such thorough vaccine screening of 100 different amastigote-expressed Leishmania genes was done in a BALB/c mouse model challenged with the *L. major* LV39 substrain using DNA vaccination [90]. A heterologous prime-boost strategy for immunising against experimental visceral leishmaniasis in dogs was successful using DNA/recombinant cysteine proteinases type I and type II [92]. We are eagerly awaiting the outcomes of additional dog experiments as well as a human study of a DNA vaccination.

12. Conclusion

Renewing hope that control is possible is the international acknowledgement of this disease's significance, coordinated by WHO programmes, particularly in India, where 70% of the world's VL burden is found. A better coordination of control programmes should be possible thanks to new techniques for early case detection. Only when the local infrastructure is developed to support the provision of healthcare will more extensive control be possible in the world's less developed areas. This attempt is hampered by the introduction of HIV, drug-resistant strains, and changes in the epidemiology of the vector. The development of new techniques and, eventually, a protective vaccine will require a sustained worldwide effort and the requisite funding.

Acknowledgements

Dr. Azhar Rafique, Doctor of Philosophy (PhD) in Zoology from University of Agriculture Faisalabad (UAF), Assistant Professor Department of Zoology at Government College University Faisalabad, Pakistan, provided help in language, edited and proofread the chapter.

Conflict of interest

The authors declare no conflict of interest.

Author details

Azhar Rafique[1*], Sayydah Surrya Sani[1], Salma Sultana[1], Tayyaba Sultana[1], Asma Ashraf[1] and Muhammad Shahid Mahmood[2]

1 Department of Zoology, Government College University Faisalabad, Punjab, Pakistan

2 Institute of Microbiology, University of Agriculture, Faisalabad, Punjab, Pakistan

*Address all correspondence to: azharrafique96@gmail.com

References

[1] Berman JD. Human leishmaniasis: Clinical, diagnostic, and chemotherapeutic developments in the last 10 years. Clinical infectious diseases. 1997;**24**:684-703. DOI: 10.1093/clind/24.4.684

[2] Kuhls K, Alam MZ, Cupolillo E, Ferreira GE, Mauricio IL, Oddone R, et al. Comparative microsatellite typing of new world Leishmania infantum reveals low heterogeneity among populations and its recent old world origin. PLoS Neglected Tropical Diseases. 2011 Jun;**5**(6):e1155. DOI: 10.1371/journal.pntd.0001155

[3] Lukes J, Mauricio IL, Schönian G, Dujardin JC, Soteriadou K, Dedet JP, et al. Evolutionary and geographical history of the Leishmania donovani complex with a revision of current taxonomy. Proceedings of the National Academy of Sciences of the United States of America. 2007;**104**(22):9375-9380. DOI: 10.1073/pnas.0703678104

[4] Maurício IL, Stothard JR, Miles MA. The strange case of Leishmania chagasi. Parasitology Today. 2000;**16**(5):188-189. DOI: 10.1016/s0169-4758(00)01637-9

[5] Momen H, Cupolillo E. Speculations on the origin and evolution of the genus Leishmania. Memórias do Instituto Oswaldo Cruz. 2000;**95**(4):583-588. DOI: 10.1590/s0074-02762000000400023

[6] Bailey MS, Lockwood DN. Cutaneous leishmaniasis. Clinics in Dermatology. 2007;**25**:203-211. DOI: 10.1016/j.clindermatol.2006.05.008

[7] Terefe Y, Afera B, Bsrat A, Syoum Z. Distribution of human leishmaniasis (VL) and its associated risk factors, in Metemma. Ethiopia. Epidemiology Research International. 2015;**22**:2015. DOI: 10.1155/2015/630812

[8] Kishore K, Kumar V, Kesari S, Dinesh DS, Kumar AJ, Das P, et al. Vector control in leishmaniasis. The Indian Journal of Medical Research. 2006;**123**:467-474

[9] Claborn DM. The biology and control of leishmaniasis vectors. Journal of Global Infectious Diseases. 2010;**2**:127-134. DOI: 10.4103/0974-777x.62866

[10] Philippe D. Leishmaniasis public health aspects and control. Elesevier science Inc. 1996;**14**:417-423. DOI: 10.1016/0738-081X(96)00057-0

[11] Schönian G, Mauricio I, Cupolillo E. Is it time to revise the nomenclature of Leishmania? Trends in parasitology. 2010;**26**:466-469. DOI: 10.1016/j.pt.2010.06.013

[12] Desjeux P. Leishmaniasis: Current situation and new perspectives. Comparative immunology, microbiology and infectious diseases. 2004;**27**:305-318. DOI: 10.1016/j.cimid.2004.03.004

[13] Hepburn NC. Cutaneous leishmaniasis: Current and future management. Expert Review of Anti-infective Therapy. 2003;**1**:563-570. DOI: 10.1586/14787210.1.4.563

[14] Martins AL, Barreto JA, Lauris JR, Martins AC. American tegumentary leishmaniasis: Correlations among immunological, histopathological and clinical parameters. Anais Brasileiros de Dermatologia. 2014;**89**:52-58. DOI: 10.1590/abd1806-4841.20142226

[15] Carvalho EM, Barral A, Costa JM, Bittencourt A, Marsden P. Clinical

and immunopathological aspects of disseminated cutaneous leishmaniasis. Acta Tropica. 1994;**56**:315-325. DOI: 10.1016/0001-706X(94)90103-1

[16] Bacellar O, Lessa H, Schriefer A, Machado P, Ribeiro de Jesus A, Dutra WO, et al. Up-regulation of Th1-type responses in mucosal leishmaniasis patients. Infection and Immunity. 2002;**70**:6734-6740. DOI: 10.1128/IAI.70.12.6734-6740.2002

[17] Hodiamont CJ, Kager PA, Bart A, de Vries HJ, van Thiel PP, Leenstra T, et al. Species-directed therapy for leishmaniasis in returning travellers: A comprehensive guide. PLoS Neglected Tropical Diseases. 2014;**8**:e2832. DOI: 10.1371/journal.pntd.0002832

[18] Savioli L, Velayudhan R. Small bite, big threat: World health day 2014. EMHJ-Eastern Mediterranean Health Journal. 2014;**20**:217-218. (World Health Organization, Regional Office for the Eastern Mediterranean. editorial)

[19] Kitchen LW, Lawrence KL, Coleman RE. The role of the United States military in the development of vector control products, including insect repellents, insecticides, and bed nets. Journal of Vector Ecology. 2009;**34**:50-61. DOI: 10.1111/j.1948-7134.2009.00007.x

[20] Bonaldo MC, Souto-Padron T, de Souza W, Goldenberg S. Cell-substrate adhesion during Trypanosoma cruzi differentiation. The Journal of Cell Biology. 1988;**106**:1349-1358. DOI: 10.1083/jcb.106.4.1349

[21] Motazedian MH, Mehrbani D, Oryan A, Asgari Q, Karamian M, Kalantari M. Life cycle of cutaneous leishmaniasis in Larestan, southern Iran. Iranian Journal of Clinical Infectious Diseases. 2006;**1**:137-143

[22] Kaye P, Scott P. Leishmaniasis: Complexity at the host–pathogen interface. Nature Reviews. Microbiology. 2011;**9**:604-615. DOI: 10.1038/nrmicro2608

[23] Sacks D, Kamhawi S. Molecular aspects of parasite-vector and vector-host interactions in leishmaniasis. Annual Review of Microbiology. 2001;**55**:453-483. DOI: 10.1146/annurev.micro.55.1.453

[24] Belkaid Y, Mendez S, Lira R, Kadambi N, Milon G, Sacks D. A natural model of Leishmania major infection reveals a prolonged "silent" phase of parasite amplification in the skin before the onset of lesion formation and immunity. Journal of Immunology. 2000;**165**:969-977. DOI: 10.4049/jimmunol.165.2.969

[25] Ribeiro-Gomes FL, Roma EH, Carneiro MB, Doria NA, Sacks DL, Peters NC. Site-dependent recruitment of inflammatory cells determines the effective dose of Leishmania major. Infection and Immunity. 2014l;**82**:2713-2727. DOI: 10.1128/IAI.01600-13

[26] Belli A, Rodriguez B, Aviles H, Harris E. Simplified polymerase chain reaction detection of new world Leishmania in clinical specimens of cutaneous leishmaniasis. AJMTH. 1998;**58**:102-109. DOI: 10.4269/ajtmh.1998.58.102

[27] de Oliveira CI, Báfica A, Oliveira F, Favali CB, Correa T, Freitas LA, et al. Clinical utility of polymerase chain reaction—Based detection of Leishmania in the diagnosis of American cutaneous Leishmaniasis. Clinical Infectious Diseases. 2003;**37**:149-153. DOI: 10.1086/379610

[28] Camargo ME, Rebonato C. Cross-reactivity in fluorescence tests

for Trypanosoma and Leishmania antibodies. A simple inhibition procedure to ensure specific results. AJMTH. 1969;**18**:500-505. DOI: 10.4269/ajtmh.1969.18.500

[29] Lugo de Yarbuh A. Studies of the leishmanin skin test positivity in cases with treatment anti-leishmania. Parasitología al día. 1997;**21**:76-80. DOI: 10.4067/S0716-07201997000300002

[30] de Lima Barros MB, Schubach A, Francesconi-do-Valle AC, Gutierrez-Galhardo MC, Schubach TM, Conceição-Silva F, et al. Positive Montenegro skin test among patients with sporotrichosis in Rio de Janeiro. Acta Tropica. 2005;**93**:41-47. DOI: 10.1016/j.actatropica.2004.09.004

[31] Reithinger R, Dujardin JC, Louzir H, Pirmez C, Alexander B, Brooker S. Cutaneous leishmaniasis. Lancet. Infec Dis. 2007;**7**:581-596. DOI: 10.1016/S1473-3099(07)70209-8

[32] Wilson ME. Leishmaniasis in Public Health and Preventive Medicine. New York: McGraw-Hill; 1998. pp. 334-336

[33] Marsden PD, Llanos-Cuentas EA, Lago EL, Cuba CC, Barreto AC, Costa JM, et al. Human mucocutaneous leishmaniasis in Três Braços, Bahia-Brazil. An area of Leishmania braziliensis braziliensis transmission. III-mucosal disease presentation and initial evolution. Revista da Sociedade Brasileira de Medicina Tropical. 1984;**17**:179-186

[34] Banerjee N. Role of I.M.a during natural calamities and national emergencies. Journal of the Indian Medical Association. 1973;**61**:477-481

[35] Rathi SK, Pandhi RK, Chopra P, Khanna N. Post-kala-azar dermal leishmaniasis: A histopathological study. Indian Journal of Dermatology, Venereology and Leprology. 2005;**71**:250-253. DOI: 10.1111/j.1365-4632.2004.01579.x

[36] Salotra P, Singh R. Challenges in the diagnosis of post kala-azar dermal leishmaniasis. The Indian Journal of Medical Research. 2006;**123**:295-310

[37] Singh N, Ramesh V, Arora VK, Bhatia A, Kubba A, Ramam M. Nodular post-kala-azar dermal leishmaniasis: A distinct histopathological entity. Journal of Cutaneous Pathology. 1998;**25**:95-99. DOI: 10.1111/j.1600-0560.1998.tb01696.x

[38] Hepburn NC, Tidman MJ, Hunter JAA. Cutaneous leishmaniasis in British troops from Belize. The British Journal of Dermatology. 1993;**128**:63-68. DOI: 10.1111/j.1365-2133.1993.tb00149.x

[39] Bryceson A. Therapy in man In: Peters W, Killick-Kendrick R edrs. The Leishmaniases in Biology and Medicine. London: Academic Press. 1987; 847-907.

[40] Herwaldt BL, Arana BA, Navin TR. The natural history of cutaneous leishmaniasis in Guatemala. The Journal of Infectious Diseases. 1992;**165**:518-527. DOI: 10.1093/infdis/165.3.518

[41] Locksley RM, Louis JA. Immunology of leishmaniasis. Current opinion in immunology. 1999;**4**:413-418. DOI: 10.1016/S0952-7915(06)80032-4

[42] Kubba R, Al-Gindan Y, El-Hassan AM, Omer AH. Clinical diagnosis of cutaneous leishmaniasis (oriental sore). J. Am. Acadm. Dermal. 1987;**16**:1183-1189. DOI: 10.1016/S0190-9622(87)70155-8

[43] Navin TR, Arana FE, de Mérida AM, Arana BA, Castillo AL, Silvers DN. Cutaneous leishmaniasis in Guatemala: Comparison of diagnostic methods.

AJTMH. 1990;**42**:36-42. DOI: 10.4269/ajtmh.1990.42.36

[44] Evans D. Handbook on Isolation, Characterisation and Cryopreservation of Leishmania. Geneva: UNDP/World Bank/WHO (TDR); 1989

[45] Herwaldt BL. Leishmaniasis. Lancet. 1999;**354**:1191-1199

[46] Weigle KA, Valderrama L, Arias AL, Santrich C, Saravia NG. Leishmanin skin test standardization and evaluation of safety, dose, storage, longevity of reaction and sensitization. AJTMH. 1991;**44**:260-271. DOI: 10.4269/ajtmh.1991.44.260

[47] El-Hassan AM, Zijlstra EE. Leishmaniasis in Sudan. Cutaneous leishmaniasis. Transactions of the Royal Society of tropical medicine and Hygiene. 2001;**1**:1-17. DOI: 10.1016/S0035-9203(01)90216-0

[48] Alvar J, Canavate C, Molina R, Moreno J, Nieto J. Canine leishmaniasis. Advances in Parasitology. 2004;**57**:1-88. DOI: 10.1016/s0065-308x(04)57001-x

[49] Baneth G, Koutinas AF, Solano-Gallego L, Bourdeau P, Ferrer L. Canine leishmaniosis–new concepts and insights on an expanding zoonosis: Part one. Trends in Parasitology. 2008;**24**:324-330. DOI: 10.1016/j.pt.2008.04.001

[50] Ciaramella P, Oliva GD, De Luna R, Ambrosio R, Cortese L, Persechino A, et al. A retrospective clinical study of canine leishmaniasis in 150 dogs naturally infected by Leishmania infantum. Veterinary record. 1997;**141**:539-543. DOI: 10.1136/vr.141.21.539

[51] Pinelli E, Killick-Kendrick R, Wagenaar J, Bernadina W, Del Real G, Ruitenberg J. Cellular and humoral immune responses in dogs experimentally and naturally infected with Leishmania infantum. Infection and immunity. 1994;**62**:229-235. DOI: 10.1128/iai.62.1.229-235.1994

[52] Dantas-Torres F. Canine leishmaniosis in South America. Parasites & Vectors. 2009;**2**:1-8. DOI: 10.1186/1756-3305-2-S1-S1

[53] Reithinger R, Davies CR. Is the domestic dog (Canis familiaris) a reservoir host of American cutaneous leishmaniasis? A critical review of the current evidence. The American Journal of Tropical Medicine and Hygiene. 1999;**61**:530-541. DOI: 10.4269/ajtmh.1999.61.530

[54] Santaella J, Ocampo CB, Saravia NG, Méndez F, Góngora R, Gomez MA, et al. Leishmania (Viannia) infection in the domestic dog in chaparral, Colombia. The American journal of tropical medicine and hygiene. 2011;**84**:674-680. DOI: 10.4269/ajtmh.2011.10-0159

[55] Spickler, Anna Rovid. Leishmaniasis. 2022. Available from: http://www.cfsph.iastate.edu/DiseaseInfo/factsheets.php

[56] Sharma NL, Mahajan VK, Kanga A, Sood A, Katoch VM, Mauricio I, et al. Localized cutaneous leishmaniasis due to Leishmania donovani and Leishmania tropica: Preliminary findings of the study of 161 new cases from a new endemic focus in Himachal Pradesh, India. The American Journal of Tropical Medicine and Hygiene. 2005;**72**:819-824. DOI: 10.4269/ajtmh.2005.72.819

[57] Hashiguchi Y, Velez LN, Villegas NV, Mimori T, Gomez EA, Kato H. Leishmaniases in Ecuador: Comprehensive review and current status. Acta tropica. 2017;**166**:299-315. DOI: 10.1016/j.actatropica.2016.11.039

[58] Garg VK, Agrawal S, Rani S, Joshi A, Agarwalla A, Das ML, et al. Post-kala-azar dermal leishmaniasis in Nepal. International Journal of Dermatology. 2001;**40**:179-184. DOI: 10.1046/j.1365-4362.2001.01198.x

[59] Rijal S, Yardley V, Chappuis F, Decuypere S, Khanal B, Singh R, et al. Antimonial treatment of visceral leishmaniasis: Are current in vitro susceptibility assays adequate for prognosis of in vivo therapy outcome? Microbes and Infection. 2007;**9**:529-535. DOI: 10.1016/j.micinf.2007.01.009

[60] Hefnawy A, Berg M, Dujardin JC, De Muylder G. Exploiting knowledge on Leishmania drug resistance to support the quest for new drugs. Trends in Parasitology. 2017;**33**(3):162-174. DOI: 10.1016/j.pt.2016.11.003

[61] Pinart M, Rueda JR, Romero GA, Pinzón-Flórez CE, Osorio-Arango K, Maia-Elkhoury AN, et al. Interventions for American cutaneous and mucocutaneous leishmaniasis. Cochrane Database of Systematic Reviews. 2020;**8**:CD004834. DOI: 10.1002/14651858.CD004834

[62] González U, Pinart M, Reveiz L, Alvar J. Interventions for Old World cutaneous leishmaniasis. Cochrane Database Syst Revs. 2008;**4**:CD005067. DOI: 10.1002/14651858.cd005067.pub3

[63] Bassiouny A, El Meshad M, Talaat M, et al. Cryosurgery in cutaneous leishmaniasis. Bri. J. Dermo. 1982;**107**:467-474. DOI: 10.1111/j.1365-2133.1982.tb00390.x

[64] Leibovici V, Aram H. Cryotherapy in acute cutaneous leishmaniasis. Int. J. Dermato. 1986;**25**:473-475. DOI: 10.1111/j.1365-4362.1986.tb03461.x

[65] Al-Majali O, Routh HB, Abuloham O, Bhowmik KR, Muhsen M, Hebeheba H. A 2-year study of liquid nitrogen therapy in cutaneous leishmaniasis. International Journal of Dermatology. 1997;**36**:460-462. DOI: 10.1046/j.1365-4362.1997.00045.x

[66] Gurei MS, Tatli N, Ozbilge H, Erel O, Seyrek A, Kocyigit A, et al. Efficacy of cryotherapy and intralesional pentostam in treatment of cutaneous leishmaniasis. Journal of the Egyptian Society of Parasitology. 2000;**30**:169-176

[67] Bern C, Joshi AB, Jha SN, Das ML, Hightower A, Thakur GD, et al. Factors associated with visceral leishmaniasis in Nepal: Bed-net use is strongly protective. The American Journal of Tropical Medicine and Hygiene. 2000;**63**:184-188. DOI: 10.4269/ajtmh.2000.63.184

[68] Tayeh A, Jalouk L, Cairncross S. Twenty years of cutaneous leishmaniasis in Aleppo, Syria. Transactions of the Royal Society of Tropical Medicine and Hygiene. 1997;**91**:657-659. DOI: 10.1016/S0035-9203(97)90509-5

[69] Maroli M, Mizzon V, Siragusa C, D'Oorazi A, Gradoni L. Evidence for an impact on the incidence of canine leishmaniasis by the mass use of deltamethrin-impregnated dog collars in southern Italy. Medical and Veterinary Entomology. 2001;**15**:358-363. DOI: 10.1046/j.0269-283x.2001.00321.x

[70] Reithinger R, Coleman PG, Alexander B, Vieira EP, Assis G, Davies CR. Are insecticide impregnated dog collars a feasible alternative to dog culling as a strategy for controlling canine visceral leishmaniasis in Brazil? International Journal for Parasitology. 2004;**34**:55-62. DOI: 10.1016/j.ijpara.2003.09.006

[71] Modabber F. Experiences with vaccines against cutaneous leishmaniasis:

Of men and mice. Parasitology. 1989;**98**:49-60. DOI: 10.1017/S0031182000072243

[72] Heyneman D. Immunology of leishmaniasis. Bulletin of the World Health Organization. 1971;**44**:499-514

[73] Lemma A, Cole L. Leishmania enriettii: Radiation effects and evaluation of radioattenuated organisms for vaccination. Experimental Parasitology. 1974;**35**:161-169. DOI: 10.1016/0014-4894(74)90019-8

[74] Streit JA, Recker TJ, Filho FG, Beverley SM, Wilson ME. Protective immunity against the protozoan Leishmania chagasi is induced by subclinical cutaneous infection with virulent but not avirulent organisms. Journal of Immunology. 2001;**166**:1921-1929. DOI: 10.4049/jimmunol.166.3.1921

[75] Sharifi I, FeKri AR, Aflatonian MR, Khamesipour A, Nadim A, Mousavi MR, et al. Randomised vaccine trial of single dose of killed Leishmania major plus BCG against anthroponotic cutaneous leishmaniasis in bam, Iran. Lancet. 1998;**351**:1540-1543. DOI: 10.1016/S0140-6736(98)09552-X

[76] Mohebali M, Khamesipour A, Mobedi I, Zarei Z, Hashemi-Fesharki R. Double-blind randomized efficacy field trial of alum precipitated autoclaved Leishmania major vaccine mixed with BCG against canine visceral leishmaniasis in Meshkin-Shahr district, I.R. Iran. Vaccine. 2004;**22**:4097-4100. DOI: 10.1016/j.vaccine.2004.03.058

[77] Lemesre JL, Holzmuller P, Cavaleyra M, Goncalves RB, Hottin G, Papierok G. Protection against experimental visceral leishmaniasis infection in dogs immunized with purified excreted secreted antigens of Leishmania infantum promastigotes. Vaccine. 2005;**23**:2825-2840. DOI: 10.1016/j.vaccine.2004.11.061

[78] Fujiwara RT, Vale AM, da Silva JC, da Costa RT, da Silva QJ, Martins Filho OA, et al. Immunogenicity in dogs of three recombinant antigens (TSA, LeIF and LmSTI1) potential vaccine candidates for canine visceral leishmaniasis. Veterinary Research. 2005;**36**:827-838. DOI: 10.1051/vetres:2005033

[79] Saraiva EM, de Figueiredo BA, Santos FN, Borja-Cabrera GP, Nico D, Souza LO, et al. The FML-vaccine (Leishmune) against canine visceral leishmaniasis: A transmission blocking vaccine. Vaccine. 2006;**24**:2423-2431. DOI: 10.1016/j.vaccine.2005.11.061

[80] Spitzer N, Jardim A, Lippert D, Olafson RW. Long-term protection of mice against Leishmania major with a synthetic peptide vaccine. Vaccine. 1999;**17**:1298-1300. DOI: 10.1016/S0264-410X(98)00363-6

[81] Wolff JA, Malone RW, Williams P, Chong W, Acsadi G, Jani A, et al. Direct gene transfer into mouse muscle in vivo. Science. 1990;**247**:1465-1468. DOI: 10.1126/science.1690918

[82] Gurunathan S, Sacks DL, Brown DR, Reiner SL, Charest H, Glaichenhaus N, et al. Vaccination with DNA encoding the immunodominant LACK parasite antigen confers immunity to mice infected with *Leishmania major*. The Journal of Experimental Medicine. 1997;**186**:1137-1147. DOI: 10.1084/jem.186.7.1137

[83] Gurunathan S, Prussin C, Sacks D, Seder R. Vaccine requirements for sustained cellular immunity to an intracellular parasitic infection. Nature Medicine. 1998;**4**:1409-1415. DOI: 10.1038/4000

[84] Belkaid Y, Hoffmann KF, Mendez S, Kamhawi S, Udey MC, Wynn TA, et al.

The role of interleukin (IL)-10 in the persistence of *Leishmania major* in the skin after healing and the therapeutic potential of anti-IL-10 receptor antibody for sterile cure. The Journal of Experimental Medicine. 2001;**194**:1497-1506. DOI: 10.1084/jem.194.10.1497

[85] Akbari O, Panjwani N, Garcia S, Tascon R, Lowrie D, Stockinger B. DNA vaccination: Transfection and activation of dendritic cells as key events for immunity. The Journal of Experimental Medicine. 1999;**189**:169-178. DOI: 10.1084/jem.189.1.169

[86] Sallusto F, Lenig D, Forster R, Lipp M, Lanzavecchia A. Two subsets of memory T lymphocytes with distinct homing potentials and effector functions. Nature. 1999;**401**:708-712. DOI: 10.1038/44385

[87] Zaph C, Uzonna J, Beverley SM, Scott P. Central memory T cells mediate long-term immunity to Leishmania major in the absence of persistent parasites. Nature Medicine. 2004;**10**:104-110. DOI: 10.1038/nm1108

[88] Stober CB, Lange UG, Roberts MT, Alcami A, Blackwell JM. IL-10 from regulatory T cells determines vaccine efficacy in murine Leishmania major infection. Journal of Immunology. 2005;**175**:2517-2524. DOI: 10.4049/jimmunol.175.4.2517

[89] Levick MP, Blackwell JM, Connor V, Coulson RM, Miles A, Smith HE, et al. An expressed sequence tag analysis of a full-length, spliced-leader cDNA library from Leishmania major promastigotes. Molecular and Biochemical Parasitology. 1996;**76**:345-348. DOI: 10.1016/0166-6851(95)02569-3

[90] Morris RV, Shoemaker CB, David JR, Lanzaro GC, Titus RG. Sandfly maxadilan exacerbates infection with Leishmania major and vaccinating against it protects against L. major infection. Journal of Immunology. 2001;**167**:5226-5230. DOI: 10.4049/jimmunol.167.9.5226

[91] Ivens AC, Peacock CS, Worthey EA, Murphy L, Aggarwal G, Berriman M, et al. The genome of the kinetoplastid parasite, *Leishmania major*. Science. 2005;**309**:436-442. DOI: 10.1126/science.1112680

[92] Rafati S, Nakhaee A, Taheri T, Taslimi Y, Darabi H, Eravani D, et al. Protective vaccination against experimental canine visceral leishmaniasis using a combination of DNA and protein immunization with cysteine proteinases type I and II of L. infantum. Vaccine. 2005;**23**:3716-3725. DOI: 10.1016/j.vaccine.2005.02.009

Chapter 6

Visceral Leishmaniasis: An Overview and Integrated Analysis of the Current Status, Geographical Distribution and Its Transmission

Kaushal Kumar Mahto, Pooja Prasad, Mohan Kumar, Intzar Ali, Vikram Vohra and Deepak Kumar Arya

Abstract

Visceral leishmaniasis (VL) is a vector–borne disease transmitted by Phlebotomine sandflies, with up to 350 million people are at risk of developing infection globally. VL has a severe influence on the impoverished and undeveloped populations among several subcontinents. Early and accurate diagnosis and treatment remain crucial to the management of VL, which still depends on vector control. The present chapter objectives are to provide an overview of visceral leishmaniasis and to raise knowledge of the most recent progress in this condition's management, treatment, and prevention. Additionally, this chapter could be helpful for comprehending the difficulties and knowledge gaps in eliminating this protozoan disease as well as for learning the planning lessons from the global management of diseases like malaria and tuberculosis.

Keywords: visceral leishmania, endemicity, surveillance, strategy, transmission control, treatment

1. Introduction

Visceral leishmaniasis (VL), also referred to as kala–azar in the Hindu vernacular, is caused by an obligate intracellular protozoon species of the genus *Leishmania* (Trypanosomatida: Trypanosomatidae) [1]. VL is transmitted by the bite of female sandfly's (Diptera, Psychodidae) [2]. VL is usually transmitted by the two species i.e. *Labrus donovani* and *L. infantum* (*L. chagasi infantum*) depending on the geographical area and generally affects internal organs such as spleen, liver, and bone marrow. It is one among the three leishmaniasis where it is life threatening, if untreated. It is mainly distributed throughout in tropical and subtropical areas and can range from asymptomatic to severe however; other forms such as cutaneous and mucosal leishmaniasis can cause substantial morbidity. VL caused by *L. infantum* mainly affects children under the age of 5 years whereas *L. donovani* infects all age groups and this may be due to malnutrition and other conditions of immunosuppression [3].

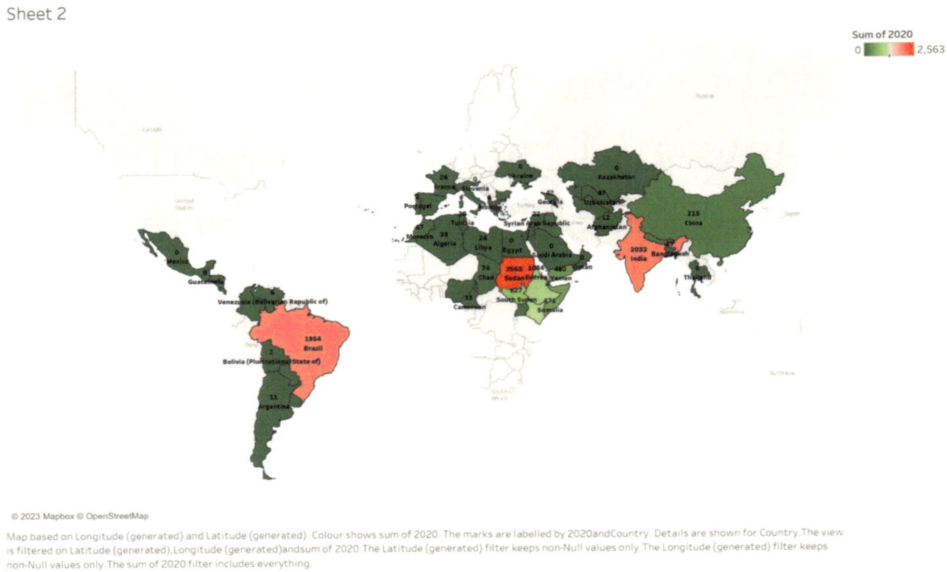

Figure 1.
Endimicity status of Visceral Leishmaniasis worldwide in the year 2020 (source: https://apps.who.int/gho/data/node.main.NTDLEISHVEND?lang=en).

The spatial distribution and the burden of VL is up surging year after year and it is now became a growing health concern worldwide. VL is thought to be an anthroponotic disease and is prevalent in several foci across Africa and the Indian subcontinent [4]. Most VL infections occur in the least developed countries and the most underdeveloped areas of middle–income countries. More than 90% of new cases reported to WHO in 2020 took place in ten countries: South Sudan, Sudan, Somalia,Yemen, Kenya, Ethiopia, Eritrea, India, Brazil, China respectively [5]. It should be possible to eradicate it with rapid case detection, treatment, and local vector control. Insecticide–impregnated materials, active case detection, and treatment are indeed the cornerstones of the current control techniques for VL.

Figure 1 represents the number of VL cases across 83 countries. In which 28 countries data are not available for the year 2020. Maximum number of VL cases were found from Sudan (2563) and most of the countries had very smaller number of cases which indicate endemicity of VL. Country like Sudan, Brazil, and India were reported maximum number of cases of VL.

2. VL and India

In India, it is estimated that 165.4 million population are at risk in 4 states viz., Bihar (33 districts, 458 blocks), Jharkhand (4 districts, 33 blocks), West Bengal (11 districts, 120 blocks), and Uttar Pradesh where kala–azar is endemic (6 districts, 22 blocks) [6] (**Table 1**). India reported 810 cases and 39 deaths in the year 2022. In order to overcome remaining obstacles in eradication of visceral leishmaniasis, India is stepping up its efforts. India has made strengthening a widely spread network of comprehensive primary health care facilities, backed by health education and social mobilization, a top priority in its national health strategy in order to attain universal

DOI: http://dx.doi.org/10.5772/intechopen.110567

Affected States/UTs	2014		2015		2016		2017		2018		2019		2020		2021(P)		2022(P)	
	C	D	C	D	C	D	C	D	C	D	C	D	C	D	C	D	C	D
Assam	1	0	1	0	0	0	0	0	0	0	—	—	—	—	—	—	—	—
Bihar	7615	10	6517	5	4773	0	4127	0	3423	0	2416	0	1427	0	893	20	547	26
Delhi*	0	0	0	0	0	0	0	0	0	0	—	—	—	—	—	—	—	—
Jharkhand	937	0	1262	0	1185	0	1358	0	752	0	541	0	424	0	275	5	187	8
Kerala	0	0	4	0	2	0	0	0	6	0	4	0	1	0	—	—	3	0
Punjab*	0	0	1	0	0	0	0	0	0	0	—	—	—	—	—	—	—	—
Sikkim	5	0	5	0	1	0	0	0	0	0	—	—	3	0	1	0	1	0
Uttarakhand	4	0	3	0	2	0	2	0	0	0	—	—	—	—	—	—	—	—
Uttar Pradesh	11	0	131	0	107	0	115	0	110	0	97	0	55	3	49	1	22	1
West Bengal	668	1	576	0	179	0	156	1	95	3	87	6	57	3	58	2	49	4
Total	9241	11	8500	5	6249	0	5758	1	4386	3	3145	6	1967	6	1276	28	810	39

*Imported.
C = Cases, D = Deaths, P = Provisional, — = Data not available.

Table 1.
VL cases and deaths in India since 2014 [7].

health coverage. For instance, there is one community health volunteer, or certified social health activist (ASHA), for every 1000 residents of a village. India now uses a 5% wettable powder formulation of alpha–cypermethrin for indoor residual spraying. Sandflies that settle on surfaces of sprayed walls are killed by it.

3. VL and suspected countries

There are 53 *Leishmania* species found worldwide in five genera (*Leishmania, Viannia, Sauroleishmania, L. enrietti complex, and Paraleishmaia*) and 31 species are known to be mammalian parasite and out of which approximately 20 different species of *Leishmania* infect animals, reservoir host includes domestic dogs and cats, wild canids and many of them can contribute to leishmaniasis in humans [8] (**Table 2**).

Major risk factors associated with VL are socioeconomic conditions, malnutrition, population mobility, environmental changes, and climate changes.

4. Growth and development of sandfly

Sandflies are tiny holometabolous insects, completes their development in four stages i.e. egg, larva, pupa and adult (**Figure 2**). Female sandflies usually lay 30–70 eggs during a single gonotrophic cycle, which are deposited in cracks and holes in the ground or in building, animal burrows and among tree roots [15]. Eggs are elongated, oval–shaped and pale at first and darkening on exposure to air. Larvae feed on dead organic material. Larvae are mainly scavengers, feeding on organic matters such as dead and decaying leaves, decomposing insets etc. The pupal stage lasts 6–13 days before the adult sandflies emerge.

Visceral leishmaniasis			
Causative parasites	**Countries (suspected)**	**Reservoir hosts**	**Reference**
L. donovani	Northeast India, Nepal, Bangladesh, Bhutan, Sri Lanka	Human	[9]
L. donovani	People's Republic of China	Unknown	[10]
L. donovani	Sudan, Ethiopia, Chad, Yemen	Human	[11]
L. donovani	Sudan, Ethiopia, Kenya, Uganda	Human	[12]
L. infantum	Med Europe, North Africa, Southwest Asia, People's Republic of China	Domestic dog, wild canids, Domestic cat	[4]
L. infantum	Latin America: not Peru or Guianas	Domestic dog, wild canids	[11]
L. tropica	Central and North Africa, Middle East, Central Asia, India	Humans, Mammals	[13]
L. martiniquensis	Martinique, Thailand	Humans, Mammals	[14]

Table 2.
Suspected countries with reservoir hosts of visceral leishmaniasis worldwide.

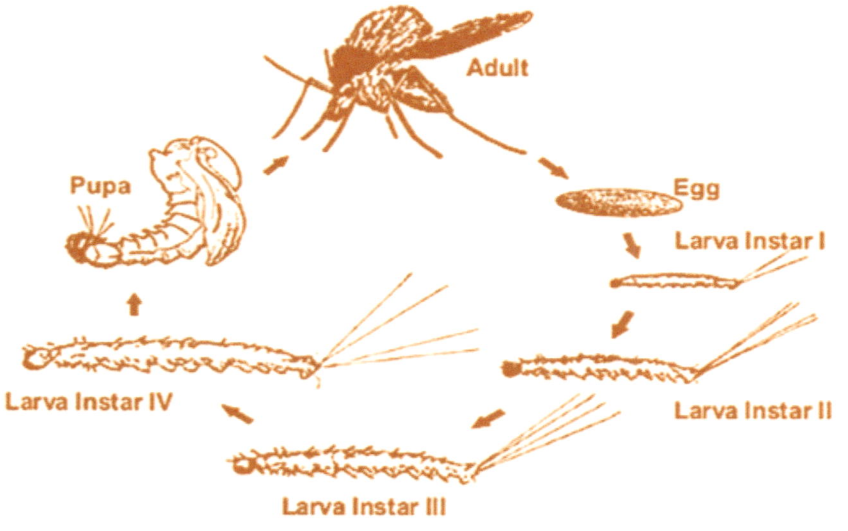

Figure 2.
A complete life cycle of sandfly.

4.1 Life cycle of *Leishmania* parasite in sandfly and host

The life cycle often involves two hosts: a phalebotomine sandfly and a vertebrate host (such as a dog or human).

According to Dawit [16], there are two evolutive forms: promastigotes for invertebrate hosts and amastigotes for vertebrate hosts.

The Following steps are involved to complete the life cycle of visceral leishmaniasis vector sandfly.

- The vector insect regurgitates the parasites when taking its blood meal.
- The host's macrophages and other mononuclear phagocytic cells phagocytose the promastigotes.
- Promastigotes transform in amastigotes and it infects other phagocytic cells.
- When sandfly ingest blood meal from another host the amastigotes are also ingested along with the blood meal.
- In the sandfly, the amastigotes develop into promastigotes in the gut, and migrate to the proboscis and the cycle continues.

The life cycle of the *Leishmania* parasite is indicated in below in **Figure 3**.

4.2 Source of infection

In both urban and rural environments, the domestic dog (*Canis familiaris*) is the primary reservoir. Foxes (*Cerdocyon thous* and *Lycalopex vetulus*) and other marsupials are among the identified reservoirs in the wild (*Didelphis spp.)* [17]. In a study

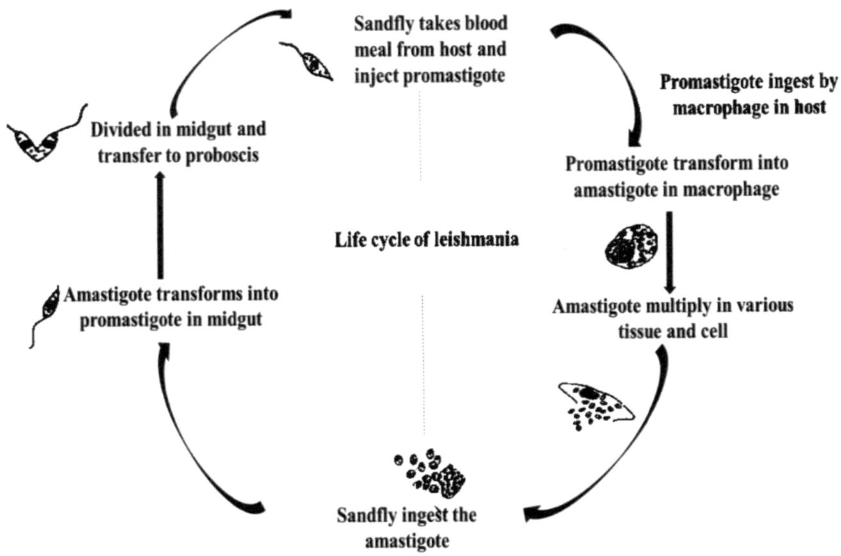

Figure 3.
Life cycle of Leishmania parasite in host and sandfly.

conducted in Iran, Gavgani *et al.* [18], the use of collars medicated with deltamethrin decreased the incidence of infection in dogs (by 54%) and children (by 43%). If a successful vaccine could be devised, vaccination of dogs would be the best approach.

4.3 Mode of transmission of VL

Leishmaniasis is caused the bite of infected female phlebotomine sand flies. The infected sand flies regurgitates the parasites infective stage (i.e., promastigotes) during blood meals According to some research studies the use of needles also contributes to parenteral and congenital transmission. No direct transfer occurs from one person to another [17]. The kind of transmission where (human—sandfly—human) involved is known as anthroponotic transmission because it occurs in some areas of the world where infected individuals are needed to keep the cycle going. For instance, the spread of *L. donovani* is anthroponotic in the Indian subcontinent (South Asia). Early diagnosis and appropriate treatment of those who are infected in these areas can act as a control strategy, whereas insufficient care can result in the development and spread of drug resistance. Spraying with residual–action pesticides and using bed nets sprayed with long–lasting insecticides may offer protection since the spread is intra– and peridomiciliary (rather than sylvatic).

4.4 Signs and symptoms

A wide spectrum of clinical manifestations, from mild to moderate to severe clinical symptoms, are present in the infection. Although it can be as long as 24 months, the incubation phase usually lasts between 2 and 6 months.

Clinical symptoms visceral infection has the following stereotypical symptoms [3, 19].

- Fever

- Weight loss
- Hepatosplenomegaly
- Pancytopenia
- Hypergammaglobulinemia

Lymphadenopathy has been reported in a few locations, most significantly Sudan and South Sudan. Other signs and symptoms include jaundice, bleeding from the mouth or nose, abdominal fluid buildup, respiratory issues, gastrointestinal issues including vomiting and diarrhea, and in extreme instances, malnutrition and lower limb edoema. In these persons deaths are eventually brought on by bacterial infection or hemorrhage.

Despite the fact that the terms kala–azar and visceral leishmaniasis are occasionally used interchangeably, kala–azar—which in Hindi is described as "black fever"—is frequently used to refer to severe (advanced) cases of visceral leishmaniasis. If untreated, visceral leishmaniasis in its most severe forms typically causes mortality, either directly from the disease or indirectly through its side effects, such as secondary Mycobacterium infection or hemorrhage.

5. Post–kala–azar dermal leishmaniasis

The dermal leishmaniasis known as post–kala–azar dermal leishmaniasis (PKDL), which typically follows visceral leishmaniasis, manifests as a macular, papular, or nodular rash on the face, upper arms, trunks, and other areas of the body [20]. The infection is said to develop in 5–10% of kala–azar patients, and it mostly affects people in East Africa and the Indian subcontinent [21] (**Table 3**). India reported 613 cases in the year 2022 where state Bihar has the heighted number of cases reported (311) followed by Jharkhand (164), West Bengal (109), and Uttar Pradesh (29). Although it can develop earlier, it often manifests 6 months to a year or more after kala–azar has purportedly been treated [6]. The Kala-azar Elimination Program (KAEP) is built around five pillars: surveillance, vector management, societal mobilization, early case detection for rapid diagnosis, treatment, and operational research. *Leishmaniasis* is thought to be transmitted by people who have PKDL.

6. VL and HIV co–infection

Due to the vulnerability of HIV–infected patients to the disease, *leishmania* and HIV confections have been a challenge to the treatment and eradication of visceral leishmaniasis. HIV and *leishmania* reinforce one another, creating serious clinical and public health challenges. Both conditions induce immune system suppression, which leads to increased death rates, exposure to drugs with higher toxicity, and more severe morbidity with fewer available treatments and higher rates of relapse. The coinfection was first identified in southern Europe in the middle of the 1980s, but is now known to exist in as many as 45 countries [6]. Brazil, Ethiopia, and the Indian state of Bihar have all recorded increased incidence. In India the subsequent cases of VL + HIV has been decreased from 9241 in 2014 to 881 cases in 2022 (**Table 4**).

Affected States	2014	2015	2016	2017	2018	2019	2020	2021 (P)	2022 (P)
	C	C	C	C	C	C	C	C	C
Bihar	119	247	542	593	731	439	351	453	311
Jharkhand	81	153	873	1211	361	281	193	208	164
West Bengal	221	255	240	166	87	51	39	59	109
Uttar Pradesh	0	0	2	12	66	50	34	50	29
Total	421	655	1657	1982	1245	821	617	770	613

C = Cases, P = Provisional.

Table 3.
PKDL situation in India since 2014 [7].

Affected States	2014	2015	2016	2017	2018	2019	2020	2021 (P)	2022(P)
	C	C	C	C	C	C	C	C	C
Bihar	NR**	40	120	181	146	121	75	74	61
Jharkhand		2	3	8	6	3	7	4	6
West Bengal		3	5	3	1	0	3	1	4
Uttar Pradesh		0	0	0	0	0	0	2	0
Total		45	128	192	153	124	85	81	71
Grand Total VL + HIV (VL)	9241	8545	6377	5950	4539	3269	2052	1357	881

*NR** = Not Reported, C = Cases, P = Provisional.*

Table 4.
VL and HIV coinfection status and situation in India from 2014 onwards [7].

Liposomal amphotericin B injections are the current treatment standard for HIV/ visceral leishmaniasis co–infection (LAmB). The new treatment regimen combines LAmB and miltefosine, an oral medication. Patients who are co–infected are at risk of developing a variety of forms of stigma and human rights issues in addition to other comorbid conditions including TB and cryptococcal meningitis. The co–infection of pulmonary TB with visceral leishmaniasis is a problem for public health in many countries. *Leishmania* infection can change the immune system's protective response to the BCG TB vaccine [22].

7. Prevention and control

No vaccination exists to protect against VL disease, however; the following are some preventive care practices for the human population to prevent vector contact [3].

1. For the human population: Avoid outdoor activities from dusk to dawn, use mosquito nets, wear protective clothes, and use insect repellents are some of the recommended personal protection measures to prevent contact with vectors.

People with clinical symptoms of the disease should receive treatment as soon as possible.

2. For vector control: Preventive methods for vector control are similar to those for cutaneous leishmaniasis in that they focus on integrated management strategies in environmental sanitation. It is recommended as a control method to apply residual insecticides safely.

3. To control the urban reservoirs: In addition to using personal protective collars on dogs, recommended preventive measures have included the placement of mosquito–proof meshes in dog kennels to keep away sandflies. In the Americas, dogs are the primary host of visceral leishmaniasis, which helps to keep the parasite alive in populated areas. As a result, it is advised to undertake canine serological surveys in areas with endemic transmission, and when a dog tests positive for the parasite, compassionate euthanasia is urged.

Drugs that are already available in the market have significant limitations in terms of cost, stability, resistance, and safety. When administered alone, they have a poor tolerability, a prolonged course of treatment, and are challenging to deliver. Most often, pentavalent antimonials are used to treat visceral leishmaniasis. However, while choosing a medicine, it is important to take the patients' clinical circumstances, the existence of any co–infections, and pregnancy in to the consideration (**Table 5**).

8. Diagnosis

Clinical: People from endemic areas are considered a high clinical suspicion of disease if they have a chronic condition, an unexplained fever, and suggestive signs and symptoms [3].

Laboratory: Following immunological and parasitological tests are performed to treat VL condition. The rapid immunochromatographic test based on recombinant rK39 antigen and the alternative Direct Agglutination Test (DAT) to confirm infection are the only immunological tests currently available at the primary level. Other levels of care also use indirect immunofluorescence (IIF) and enzyme immunoassay (ELISA) [24, 25].

Drugs	Issues
Pentavalent antimonials (Sodium stibogluconate and meglumine antimoniate)	Resistance, toxicity in HIV coinfection higher costs
Amphotericin	Need for intravenous infusion dose–limiting toxicity
Lipid–associated liposomal	Cost and cost effectiveness reportedly more effective and less toxic, intolerance and HIV–*leishmania* co–infection.
Colloidal dispersion	Should be administered in hospital setting
Lipid complex Pentamidine	Increasing unresponsiveness (in India)

Table 5.
Anti–leishmanial drugs in current use and issues [23].

Figure 4.
Methods used for the diagnosis of Visceral leishmaniasis.

By conducting direct examinations or isolating parasites in culture, parasitological procedures can identify parasites in affected organs, primarily the bone marrow (in vitro). PCR could also be used to find out *Leishmania* infection (**Figure 4**).

9. Planning lessons from the global management of diseases like malaria and tuberculosis

- Monitoring disease trends, evaluating the effectiveness of control measures, and supporting national leishmaniasis control programmes financially and technically.
- Encouraging the development and use of leishmaniasis prevention measures, including as vaccinations, diagnostic tools, and safe, efficient, and affordable therapeutics.
- Creating norms and policies based on scientific evidence, ensuring their application, and preventing and controlling leishmaniasis Healthcare professionals should receive pre- and post-assessment training.
- Digital data reporting employing mobile application surveillance tools can be utilized for effective control and successful programme in the system.

- To decide whether to use alternative therapies, it is necessary to rigorously and continuously assess resistance. Insecticides and next-generation drugs should both be the subject of concurrent study.

10. Conclusion

In high–burden countries, early detection, management, and treatment results should be improved by good community knowledge of the condition, the availability of resources, the development of clinical and health professionals' capacities, and reliable surveillance data. In summary, there is a strong need for continuous investment in VL diagnostic, treatment, and prevention, even though much can be accomplished with the already available tools and techniques. All of the existing drugs have one or more limitations, thus further funding for drug development is still desperately necessary to fill the pipeline with novel drugs. The investigation of combination therapy using already available medicines continues to be a top goal in the meantime. In areas where *L. infantum* is prevalent, novel approaches to reduce the animal reservoir (such as dog collars) or minimize human infection are needed (such as insecticide impregnated bed nets or blankets).

Acknowledgements

Authors are gratefully acknowledging National Institute of Tuberculosis and Respiratory Diseases, New Delhi, India for providing the suitable facilities that help us to draft the proposed chapter.

Conflict of interest

The authors declare that they have no known competing financial interests or personal ties that would seem to have influenced the work presented in this study.

Author details

Kaushal Kumar Mahto[1*], Pooja Prasad[2,3], Mohan Kumar[4], Intzar Ali[5], Vikram Vohra[1] and Deepak Kumar Arya[3]

1 National Institute of Tuberculosis and Respiratory Diseases, New Delhi, India

2 National Institute of Malaria Research, New Delhi, India

3 Department of Zoology, Kumaun University, Nainital, Uttarakhand, India

4 All India Institute of Medical Sciences, New Delhi, India

5 Hamdard Institute Medical Sciences and Research, New Delhi, India

*Address all correspondence to: kausha63_sit@jnu.ac.in

References

[1] WHO. Control of the leishmaniasis. In: Report of a Meeting of the WHO Expert Committee on the Control of Leishmaniasis. Geneva; 2010

[2] Ready PD. Biology of phlebotomine sand flies as vectors of disease agents. Annual Review of Entomology. 2013;**58**:227-250. DOI: 10.1146/annurev–ento–120811-153557

[3] Chappuis F et al. Visceral leishmaniasis: What are the needs for diagnosis, treatment and control? Nature Reviews. Microbiology. 2007;**5**(11):873-882. DOI: 10.1038/nrmicro1748

[4] Ready PD. Epidemiology of visceral leishmaniasis. Clinical Epidemiology. 2014;**6**:147-154. DOI: 10.2147/CLEP.S44267

[5] WHO. 2021. Available from: who.int/data/gho/data/indicators/indicator-details/GHO/status-of endemicity-of-visceral-leishmaniasis

[6] WHO. Leishmaniasis. 2022. Available from: https://www.who.int/news–room/fact–sheets/detail/leishmaniasis

[7] NVBDCP2023. Available from: https://nvbdcp.gov.in/index1.php?lang=1&level=1&sublinkid=5774&lid=3692

[8] Akhoundi M et al. A historical overview of the classification, evolution, and dispersion of Leishmania parasites and Sandflies. PLoS Neglected Tropical Diseases. 2016;**10**:e0004349. DOI: 10.1371/journal.pntd.0004349

[9] Thakur L et al. Atypical leishmaniasis: A global perspective with emphasis on the Indian subcontinent. PLoS Neglected Tropical Diseases. 2018;**12**(9):e0006659. DOI: 10.1371/journal.pntd.0006659

[10] Zhao-Rong L et al. Visceral leishmaniasis in China: An endemic disease under control. Clinical Microbiology Reviews. 2015;**28**(4):987-1004. DOI: 10.1128/CMR.00080-14

[11] Inceboz T. Epidemiology and ecology of Leishmaniasis. In: Current Topics in Neglected Tropical Diseases. London, UK: IntechOpen; 2019. DOI: 10.5772/intechopen.86359

[12] Gadisa E et al. Eco–epidemiology of visceral leishmaniasis in Ethiopia. Parasites & Vectors. 2015. Available from: https://bit.ly/3zckvBd

[13] Akhoundi M et al. Geographical distribution of *Leishmania* species of human cutaneous Leishmaniasis in Fars Province, Southern Iran. Iranian Journal of Parasitology. 2013;**8**:85-91

[14] Desbois N et al. *Leishmania (Leishmania) martiniquensis* n. sp. (Kinetoplastida: Trypanosomatidae), description of the parasite responsible for cutaneous leishmaniasis in Martinique Island (French West Indies). Parasite. 2014;**21**:12. DOI: 10.1051/parasite/2014011

[15] Service M. Phlebotomine sand–flies. In: Service M, editor. Medical Entomology for Students. 4th ed. Vol. 18. Cambridge: Cambridge University Press; 2008

[16] Dawit G, Girma Z, Simenew K. A review on biology, epidemiology and public health significance of Leishmaniasis. Journal of Bacteriology Parasitology. 2013;**4**:166. DOI: 10.4172/2155-9597-1000166

[17] PAHO. Available from: https://www3.paho.org/hq/index.

php?option=com_content&view=a
rticle&id=6420:2012-leishmaniasis-visceral&Itemid=39347&lang=en#gsc.tab=0

[18] Gavgani ASM, Hodjati MH, Mohite H, Davies CR. Effect of insecticide–impregnated dog collars on incidence of zoonotic visceral leishmaniasis in Iranian children: A matched cluster randomized trial. Lancet. 2002;**360**(9330):374-379. DOI: 10.1016/s0140-6736(02)09609-5

[19] CDC. Available from: https://www.cdc.gov/parasites/leishmaniasis/gen_info/faqs.html

[20] Zijlstra EE. The immunology of post–kala–azar dermal leishmaniasis (PKDL). Parasites & Vectors. 2016;**9**(1):464. DOI: 10.1186/s13071-016-1721-0

[21] Kumar R et al. Para–kala–azar dermal Leishmaniasis cases in Indian subcontinent – A case series. Pathogens Global Health. 2016;**110**(7-8):326-329. DOI: 10.1080/20477724.2016.1258163

[22] WHO. Available from: https://www.who.int/news–room/fact–sheets/detail/leishmaniasis

[23] Jaiprakash et al. Anti leishmanial drug discovery: Comprehensive review of the last 10 years. RSC Advances. 2015;**5**. DOI: 10.1039/C5RA02669E

[24] Srivastava P, Dayama A, Mehrotra S, Sundar S. Diagnosis of visceral leishmaniasis. Transactions of the Royal Society of Tropical Medicine and Hygiene. 2011;**105**(1):1-6. DOI: 10.1016/j.trstmh.2010.09.006

[25] Thakur S, Joshi J, Kaur S. Leishmaniasis diagnosis: An update on the use of parasitological, immunological and molecular methods. Journal of Parasitic Diseases. 2020;**44**(2):253-272. DOI: 10.1007/s12639-020-01212–w

Chapter 7

From Biology to Disease: Importance of Species-Specific *Leishmania* Antigens from the Subgenera *Viannia* (*L. braziliensis*) and *Leishmania* (*L. amazonensis*) in the Pathogenesis of American Cutaneous Leishmaniasis

Fernando T. Silveira, Marliane B. Campos, Silvia F. Müller, Patrícia K. Ramos, Luciana V. Lima, Thiago V. dos Santos, Claudia Maria Gomes, Márcia D. Laurenti, Vania Lucia da Matta and Carlos Eduardo Corbett

Abstract

American cutaneous leishmaniasis (ACL) is one of the most complex parasitic diseases from a clinical-immunopathological point of view due to the great heterogeneity of *Leishmania* species responsible for the disease. Currently, fifteen *Leishmania* species of the subgenera *Leishmania*, *Viannia* and *Mundinia* may give rise to ACL in Latin America. In Brazil, seven species are associated to the disease, but *L. (V.) braziliensis* and *L. (L.) amazonensis* stand out for producing the broadest clinical-immunopathological spectrum: localized cutaneous leishmaniasis [LCL: $DTH^{+/++}$], borderline disseminated cutaneous leishmaniasis [BDCL: $DTH^{+/-}$], mucocutaneous or mucosal leishmaniasis [MCL/ML: DTH^{++++}], and anergic diffuse cutaneous leishmaniasis [ADCL: DTH^{-}]. Although human genetic profile plays important factor in the immunopathogenesis of ACL, it deserves to be highlighted the crucial role of species-specific antigens of *L. (V.) braziliensis* and *L. (L.) amazonensis* [lipophosphoglycans, phosphatidylserine, proteophosphoglycans, glycoprotein-63 and CD200 – a macrophage activation inhibitor molecule] in the modulation of T-cell immune response ($CD4^{+}/CD8^{+}$) that will define the infection evolution.

Keywords: American cutaneous leishmaniasis, *Leishmania (V.) braziliensis*, *Leishmania (L.) amazonensis*, species-specific antigens, pathogenesis

1. Introduction

American cutaneous leishmaniasis (ACL) is an infectious, non-contagious, chronic disease caused by fifteen well recognized species of protozoan parasites of the genus *Leishmania* Ross 1903, widely distributed throughout Latin America. The disease is characterized by involvement of the cutaneous and/or mucosal tissue, where leishmaniotic ulcer represents the most usual clinical manifestation of ACL. However, other types of cutaneous lesions can be seen, such as: papule, tubercle, nodule, infiltrated plaque and verrucous or keloid-like vegetating lesion, among others. The mucosal lesion can be installed in the nose, mouth or pharynx alone, or even simultaneously, however, it is more frequent in the nasal mucosa, giving it a granulomatous ulcerative aspect, which, depending on its depth, can reach the cartilaginous tissue causing perforation of the nasal septum [1–10].

In Brazil, the leishmaniotic nature of the skin and/or mucosal lesions of ACL were only confirmed, for the first time, in 1909, by Lindenberg, who found amastigote forms of *Leishmania* in smears made from the skin and/or mucosal lesions of patients coming from the interior of the State of São Paulo [southeastern Brazil]. Nevertheless, it was Vianna who described the new parasite, *Leishmania braziliensis* Vianna 1911, the first *Leishmania* species recognized in the New World. From then on, the etiology of ACL referred to as "Bauru's ulcer", "brave wound" or "tapir nose" was finally characterized [11, 12].

Until the early 1970s, all clinical forms of ACL in Brazil were attributed to a single *Leishmania* species, *Leishmania braziliensis* [13]. However, research carried out in the Brazilian Amazon pointed to the presence of a new *Leishmania* species responsible for some cases of the "anergic diffuse cutaneous form" named *Leishmania mexicana amazonensis* Lainson & Shaw 1972 [=*Leishmania (Leishmania) amazonensis* Lainson & Shaw 1987] [14]. Later, as new *Leishmania* species were identified in association with ACL [15, 16], great heterogeneity was noticed among these parasites which motivated a new classification of these protozoan parasites of the genus *Leishmania*, whose taxonomic character of greater expression was the biological behavior of the parasite within the gut of the sandfly vectors, represented by the terms: "hypopylaria, peripylaria and suprapilaria" [14]. Thus, these authors proposed the taxonomic classification of all *Leishmania* species into two subgenera: *Leishmania* Saf'janova 1982, and *Viannia* Lainson & Shaw 1987, in which the species that were previously allocated within the "*mexicana*" complex [suprapylaria = with development from the midgut to the foregut] were then allocated within the subgenus *Leishmania* [the same name of the genus], while those that were previously allocated within the "*braziliensis*" complex [peripylaria = with development from the hindgut (pylorus) to the foregut] were allocated within the subgenus *Viannia*.

Currently, these two subgenera [*Leishmania* and *Viannia*] include all seven *Leishmania* species and a hybrid parasite responsible for ACL in Brazil, mainly in the Brazilian Amazon, such as: *L. (V.) braziliensis*, *L. (V.) guyanensis*, *L. (V.) shawi*, *L. (V.) lainsoni*, *L. (V.) lindenbergi*, *L. (V.) naiffi*, *L. (L.) amazonensis* and *L. (V.) guyanensis*/*L. (V.) shawi* [9–12, 17, 18]. Nevertheless, it is important to note that, more recently, a new molecular approach of the V7V8 SSU rRNA, Hsp70, and gGAPDH genes enabled the description of a new subgenus named *Mundinia* Shaw, Camargo & Teixeira 2016 [19], where *L. (M.) martiniquensis* Desbois et al. 2014 is classified, an ACL agent in the Caribbean region (**Table 1**).

Viannia Lainson & Shaw 1987	Subgenera *Leishmania* Ross 1903	*Mundinia* Shaw et al. 2016
*L. (V.) braziliensis** Vianna 1911	*L. (L.) mexicana* Biagi 1953	*L. (M.) martiniquensis* Desbois et al. 2014
L. (V.) peruviana Velez 1913	*L. (L.) pifanoi* Medina/Romero 1959	
*L. (V.) guyanensis** Floch 1954	*L. (L.) amazonensis**Lainson/Shaw 1972	
L. (V.) panamensis Lainson/Shaw 1972	*L. (L.) garnhami* Scorza et al. 1979	
*L. (V.) lainsoni** Silveira et al. 1987	*L. (L.) venezuelensis* Bonfante-Garrido 1980	
*L. (V.) naiffi** Lainson/Shaw 1989	*L. (L.) waltoni* Shaw et al. 2015	
*L. (V.) shawi** Lainson et al. 1989		
*L. (V.) lindenbergi** Silveira et al. 2002 *L. (V.) utingensis* ‡ Braga et al. 2003		

**Etiological agents of American cutaneous leishmaniasis [ACL] in Brazil.*
‡Unique record of human infection by molecular biology [ITS-1 PCR].

Table 1.
Taxonomic classification of Leishmania parasites responsible for the etiology of American cutaneous leishmaniasis [ACL].

In Brazil, from an eco-epidemiological point of view, ACL is regarded as a primary zoonotic infection of wild mammals, such as rodents, marsupials, edentates and primates, whose transmission of the parasites between their mammalian hosts is carried out by the infected females of different species of sandfly vectors [Diptera: Psychodidae: Phlebotominae] [11, 12, 20–24]. In this country, the disease represents an important public health problem, mainly in the North, Northeast and Midwest regions, where ecological, climatic and socio-economic factors have contributed to the formation and maintenance of large foci of the disease [9, 10, 25, 26]. In the North region, that is, in the Brazilian Amazon, it deserves special attention not only due to its high casuistry, but also due to its great heterogeneity of *Leishmania* species, mainly in the State of Pará, where *L. (V.) braziliensis* and *L. (L.) amazonensis* have high pathogenic potential to give rise to one of the most complex parasitic diseases.

In this sense, it is worth mentioning that, together, *L. (V.) braziliensis* and *L. (L.) amazonensis* are responsible for the broadest clinical-immunopathological spectrum of ACL already known, consisting of four clinical forms: localized cutaneous leishmaniasis [LCL], the most common form placed at the center of this spectrum and supported by a well-balanced T-cell immune response, including the delayed-type hypersensitivity reaction (DTH), mainly in cases by *L. (V.) braziliensis* [$DTH^{+/++}$/$CD4^+ < CD8^+$/Th1 > Th2], but with a high prevalence rate [≥80%] suppression of the T-cell hypersensitivity response in cases by *L. (L.) amazonensis* [$DTH^{\pm/-}$/$CD4^+ < CD8^+$/Th1 > Th2]; mucocutaneous or mucosal leishmanaisis [MCL/ML], the extreme expression of the T-cell hypersensitivity response linked to a strong species-specific $CD4^+$/Th1-type immune response against to infection [DTH^{++++}/$CD4^+ > CD8^+$/Th1 > Th2]; anergic diffuse cutaneous leishmaniasis [ADCL], the extreme expression of the T-cell hyposensitivity response linked to a strong species-specific $CD4^+$/Th2-type immune response against to infection [DTH^-/$CD4^+ < CD8^+$/Th1 < Th2]; and borderline disseminated cutaneous leishmaniasis [BDCL], an

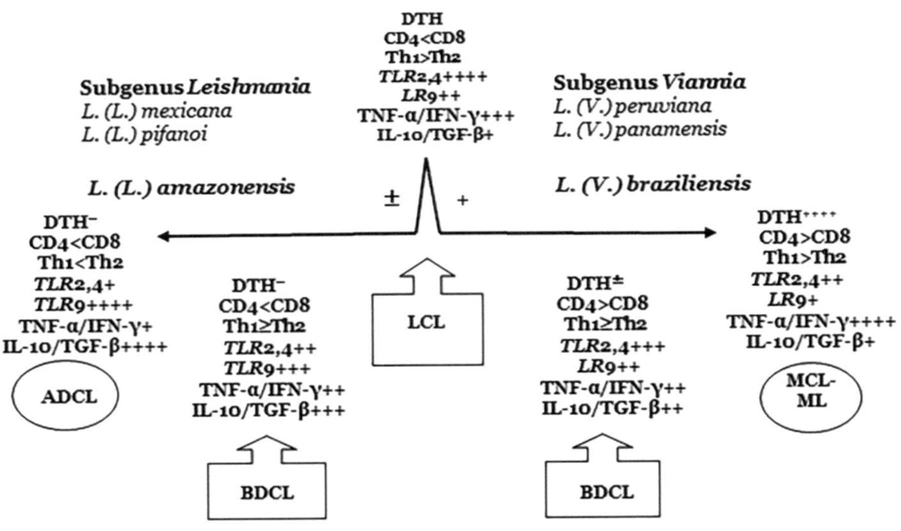

Figure 1.
Some clinical and immunopathological features of the ACL spectrum caused by L. (V.) braziliensis *and* L. (L.) amazonensis *in Brazil. Anergic DCL = diffuse cutaneous leishmaniasis; BDCL = borderline disseminated cutaneous leishmaniasis; LCL = localized cutaneous leishmanaisis; ML = mucosal leishmaniasis; DTH = delayed-type hypersensitivity reaction; CD4 = lymphocyte TCD4; CD8 = lymphocyte TCD8; Th1 = Th1-type immune response; Th2 = Th2-type immune response; TLR = tool-like receptors 2, 4 and 9; TNF-α = tumor necrosis factor-alpha; IFN-γ = interferon-gamma; IL-10 = interleukin-10; TGF-β = tumor growth factor-beta. (−) no reaction; (±) borderline reaction; (+) weak reaction; (++) moderate reaction; (+++) strong reaction; and (++++) exacerbated reaction.*

intermediary form between the central LCL and the two polar MCL/ML and ADCL forms, which is distinguished by a partial or incomplete suppression of the T-cell immune response more evident in those cases due to *L. (L.) amazonensis* [DTH^{-}/$CD4^{+} < CD8^{+}$/Th1 ≥ Th2] than those due to *L. (V.) braziliensis* [$DTH^{\pm}$/$CD4^{+} > CD8^{+}$/Th1 ≥ Th2] (**Figure 1**) [10, 18, 27–30].

That said, one could question the real significance of the taxonomic classification of the genus *Leishmania* proposed by Lainson & Shaw [14] based on a merely biological taxonomic character in relation to the pathogenesis of ACL. In fact, perhaps not even the authors themselves at that time realized how significant that taxonomic classification would have in relation to the pathogenesis of ACL. However, the immunopathogenic competence of those *Leishmania* species of the subgenera *Leishmania* and *Viannia* play in the pathogenesis of ACL is now better understood. In this way, it has been shown a clear dichotomy on the interaction between *L. (L.) amazonensis* and *L. (V.) braziliensis* with human T-cell immune response; while *L. (L.) amazonensis* shows a clear tendency to lead infection from the LCL, a persistent suppressed T-cell hypersensitivity response [$DTH^{+/-}$] clinical form in the center of the spectrum, towards the ADCL at the T-cell hyposensitivity pole [DTH^{-}] and with a marked $CD4^{+}$/Th2-type immune response, *L. (V.) braziliensis* shows an opposite trend from the well-balanced form of the T-cell hypersensitivity [$DTH^{+/++}$] in the center of the spectrum, leading infection towards the MCL/ML in the T-cell hypersensitivity pole [DTH^{++++}] and with a prominent $CD4^{+}$/Th1-type immune response [18, 27, 29–33]. More than that, there is now some revealing evidence regarding the importance of some species-specific *Leishmania* antigens of the subgenera *Leishmania* and *Viannia*, represented by a dense layer of glycoconjugate molecules also known as virulence factors, such as, lipophosphoglycans [LPG], glucoinositolphospholipids [GIPLs], proteophosphoglycans [PPGs], glycoprotein-63 [GP-63], phosphatidylserine [PS], and CD200 [macrophage activation inhibitor molecule], playing a crucial role in modulating T-cell immune response against infection, which will be the reason for a more complete approach later in this chapter [18, 34–36].

2. Clinical and immunopathological spectrum of ACL due to *L. (V.) braziliensis* and *L. (L.) amazonensis*

Firstly, what reasons would there be today to highlight the importance of *L. (V.) braziliensis* and *L. (L.) amazonensis* as the species of greatest medical interest regarding the clinical-immunopathological spectrum of ACL in Brazil, mainly in the Brazilian Amazon. Without a doubt, based on what has already been said above and also taking in account that *L. (V.) braziliensis* is historically the first *Leishmania* species described in the New World, where it shows the widest geographic distribution in Latin America, from Central America [where it has been identified as an agent of ACL in Honduras, Belize, Costa Rica, Ecuador, Guatemala, Mexico, Nicaragua, and Panama] to South America [where it is strongly associated to ACL, mainly in Brazil, but also in Colombia, Venezuela, Peru, Bolivia, Paraguay, and Argentina], it seems fair enough to regard *L. (V.) braziliensis* as the main *Leishmania* species in the context of ACL [10–12, 37].

On the other hand, alongside *L. (V.) braziliensis* in Brazil, there seems to be no doubt that *L. (L.) amazonensis* occupies the second position in the ranking of the most medically important *Leishmania* species in reason of its capacity to develop the ADCL, the extreme expression of T-cell hyposensitivity pole [DTH^-], which although it occurs with low frequency is capable of reproducing severe [nodules and infiltrated plaques] and disseminated [sparing only the scalp] chronic and incurable skin lesions in patients with strong suppression of the T-cell immune response [$CD4^+ < CD8^+$/Th1 < Th2/IL-10 and TGF-β > TNF-α and IFN-γ] [18, 27, 29, 30, 38].

Based on what was said above, it does not seem inappropriate to assume that *L. (V.) braziliensis* and *L. (L.) amazonensis* are recognized as the two *Leishmania* species with the greatest pathogenic potential for reproducing the clinical-immunopathological spectrum of ACL in Brazil: LCL and BDCL by both *Leishmania* species, while MCL/ML by *L. (V.) braziliensis* alone, as well as ADCL by *L. (L.) amazonensis* alone, which, from now on, reinforces the crucial role of the species-specific antigens also known as virulence factors [LPGs, GIPLs, PPGs, GP-63, PS and CD200] of these *Leishmania* species in the orchestration of the clinical-immunopathological spectrum of ACL. In other words, while species-specific *L. (V.) braziliensis*-antigens modulate a strong pro-inflammatory response in MCL/ML with production of high levels of major pro-inflammatory cytokines [IFN-γ/TNF-α], species-specific *L. (L.) amazonensis*-antigens modulate a strong anti-inflammatory response in ADCL with production of high levels of major regulatory cytokines [IL-10/TGF-β], even though higher ($P < 0.05$) antigen-presenting cells densities [dendritic Langerhans cell (LC): CD1a– $MHCII^+$, $Langerin^+$, and dermal dendritic cell (dDC): CD11c–$MHCII^+$, $Langerin^-$ and $MHCI^+$, $Langerin^+$] have been shown in ADCL clinical form, in the T cell hyposensitivity pole [DTH^-], compared to those in LCL clinical form, a well-balanced T-cell hypersensitivity form [$DTH^{+/++}$] in the ACL spectrum, evidencing a crucial role of these species-specific antigens of *L. (L.) amazonensis* and *L. (V.) braziliensis* in modulating T-cell immune response ($CD4^+$/$CD8^+$) against infection [18, 27, 29, 30, 39–41].

Nevertheless, one could question the non-inclusion of *L. (V.) guyanensis*, alongside *L. (V.) braziliensis*, among the species with pathogenic potential to cause MCL/ML in Brazil. In fact, recent evidence has shown *L. (V.) guyanensis* as a causative agent of MCL/ML in the Brazilian Amazon [42], principally in the north bank of the Amazon River in the State of Amazonas, which undoubtedly limits its importance as a causative agent of MCL/ML not only in the Brazilian territory, but in the Brazilian Amazon as well. In the State of Pará, for example, where MCL/ML patients are regularly present in the outpatient clinic of the "Ralph Lainson leishmaniasis laboratory" at the Evandro Chagas Institute [Science, Technology and Innovation Secretariat, Ministry of Health, Brazil] in the municipality of Ananindeua, whose demand for patients comes from all regions of the State of Pará and to a lesser extent from the States of Amapá, Maranhão, Tocantins and Mato Grosso, the only leishmanine parasite that has been identified from MCL/ML in the last forty years is *L. (V.) braziliensis*, which leaves no doubt as to its importance as the main agent of MCL/ML in Brazil [9–12, 14, 18, 27, 43, 44].

2.1 Localized cutaneous leishmaniasis (LCL) due to *L. (V.) braziliensis* and *L. (L.) amazonensis*

In Brazil, LCL is, without a doubt, the most common clinical manifestation of ACL, having as causal agent any of the seven *Leishmania* species and the hybrid parasite [*L. (V.) guyanensis/L. (V.) shawi*] mentioned above, that is, six species of the subgenus *Viannia*, including the species of greatest medical interest like *L. (V.) braziliensis*, and one of the subgenus *Leishmania*, with *L. (L.) amazonensis* being the only species associated with ACL in Brazil. However, it is necessary to record that, although the medical importance of these two species is well recognized, the association of *L. (V.) braziliensis* with the disease is much greater than that of *L. (L.) amazonensis*, which is explained due to the higher anthropophilic potential of the main sandfly vectors of *L. (V.) braziliensis*, such as, *Lutzomyia intermedia*, *Lut. whitmani*, *Lut. migonei*, *Psychodopygus wellcomei*, *Psy. complexus* and *Psy. davisi* than that of *Lut. flaviscutellata*, the principal sandfly vector of *L. (L.) amazonensis* in Brazil [9, 10, 45–48].

From a clinical point of view, however, it is important to say that, although the typical ulcerated skin lesion, with a raised and infiltrated edge and a granulomatous background, represents the major clinical sign most frequently observed in patients with LCL caused by *L. (V.) braziliensis* or any other species of the subgenus *Viannia*, especially after one to two months of disease progression (**Figure 2A**), the same cannot be said in relation to the skin lesion in patients with LCL caused by *L. (L.) amazonensis* as, in most cases, it is represented by a nodule or plaque with a highly infiltrated base and edge, with a thickened appearance, evidencing discrete ulceration or even an exulceration on the surface (**Figure 2E,F**), without that typical ulcerative character seen in the skin lesion caused by *L. (V.) braziliensis*. This clinical difference in the skin lesion of LCL between *L. (V.) braziliensis* and *L. (L.) amazonensis* clearly reflects distinct physiopathogenic mechanisms on the interaction of species-specific antigens of these *Leishmania* species with human T-cell immune response [$CD4^+/CD8^+$]. On the one hand, while *L. (V.) braziliensis* is able to induce a moderate to overt $DTH^{(+/++)}$ in LCL patients, on the other, *L. (L.) amazonensis* interacts in contrary to this by suppressing the $DTH^{(\pm/-)}$ in more than 80% of LCL patients, which might explain the greater severity of dermal necrosis and subsequent ulceration seen in the skin lesion induced by *L. (V.) braziliensis*, since dermal necrosis represents a defense mechanism in the elimination of the parasite strongly related to DTH response [49, 50]. On the contrary, in the skin lesion induced by *L. (L.) amazonensis* dermal necrosis is a more discrete phenomenon due to the suppression of DTH$^{(\pm/-)}$, resulting in a heavy parasite load on the vacuolated macrophages and lesser intense ulcerative process compared to that seen in the skin lesion of LCL by *L. (V.) braziliensis* [6, 18, 27, 29, 51–53].

Another clinical feature that deserves to be highlighted here refers to the polymorphism of the skin lesions that can be seen in patients with LCL caused by *L. (V.) braziliensis*, where not only the typical ulcerated skin lesion can be seen but also papule-like lesions, tubercle, nodule, infiltrated plaque and verrucous or keloid vegetating lesion (**Figure 2B–D**), while in patients with LCL caused by *L. (L.) amazonensis* the skin lesions do not go beyond the nodule or infiltrated plaque with a highly infiltrated base and edge, with a thickened appearance, evidencing discrete ulceration or even an exulceration on the surface (**Figure 2E,F**). However, in general, these clinical features are not optimized by many authors in Brazil, perhaps due to the low frequency of patients with LCL caused by *L. (L.) amazonensis* outside the Brazilian Amazon or the lack of a specific diagnosis of the etiologic agent involved. The fact is that, in the State of Pará, in the Brazilian Amazon, a regular number (194) of patients with LCL caused by *L. (L.) amazonensis* have been observed in the last forty years in the outpatient clinic of the "Ralph Lainson leishmaniasis laboratory" at the Evandro Chagas Institute [Science, Technology and Innovation Secretariat, Ministry of Health, Brazil] (Silveira, personal observation), more precisely from the northeast of the State of Pará, which has allowed a significant gain regarding the clinical, histopathological and immunopathological understanding of LCL caused by *L. (L.) amazonensis* [18, 27, 29, 38, 51–55].

However, before the establishment of *Leishmania*-infection, either by *L. (V.) braziliensis* or *L. (L.) amazonensis* and the subsequent development of the skin lesion, that is, of the LCL itself, one cannot fail to remember the crucial role of those species-specific antigens of *L. (V.) braziliensis* and *L. (L.) amazonensis* play during the two major events in the development of infection: first, the event that determines *Leishmania*-infection, represented by the interaction parasite-macrophage or parasite-neutrophil-macrophage; and second, the event that determines the evolution of *Leishmania*-infection, represented by the interaction parasite-dermal dendritic cell and/or parasite-Langerhans dendritic cell with human T-cell immune response [$CD4^+/CD8^+$] [9, 10, 18, 29, 32, 33, 40, 56, 57]. The recognition of these two events markedly influenced by the species-specific antigens of *L. (V.) braziliensis* and *L. (L.) amazonensis* is of great interest for a better understanding of the clinical-immunopathological spectrum of ACL caused by these *Leishmania* species.

Thus, after the establishment of the infection in the macrophage, some individuals who show effective activation of the innate immune response via IFN-γ and TNF-α are able to neutralize the infection [asymptomatic], while the majority develop different degrees of susceptibility [symptomatic], which may give rise to the central LCL form of the clinical-immunopathological spectrum of ACL that, depending on the species-specific *Leishmania*-antigens involved, such as, *L. (V.) braziliensis* or *L. (L.) amazonensis*, may present different clinical, histopathological and immunopathological profiles of infection [29, 32, 58].

In LCL caused by *L. (V.) braziliensis*, it has already been seen above that the predominant skin lesion is the typical leishmaniotic ulcer, which can generally be recognized after one to two months of disease progression, revealing, in most cases,

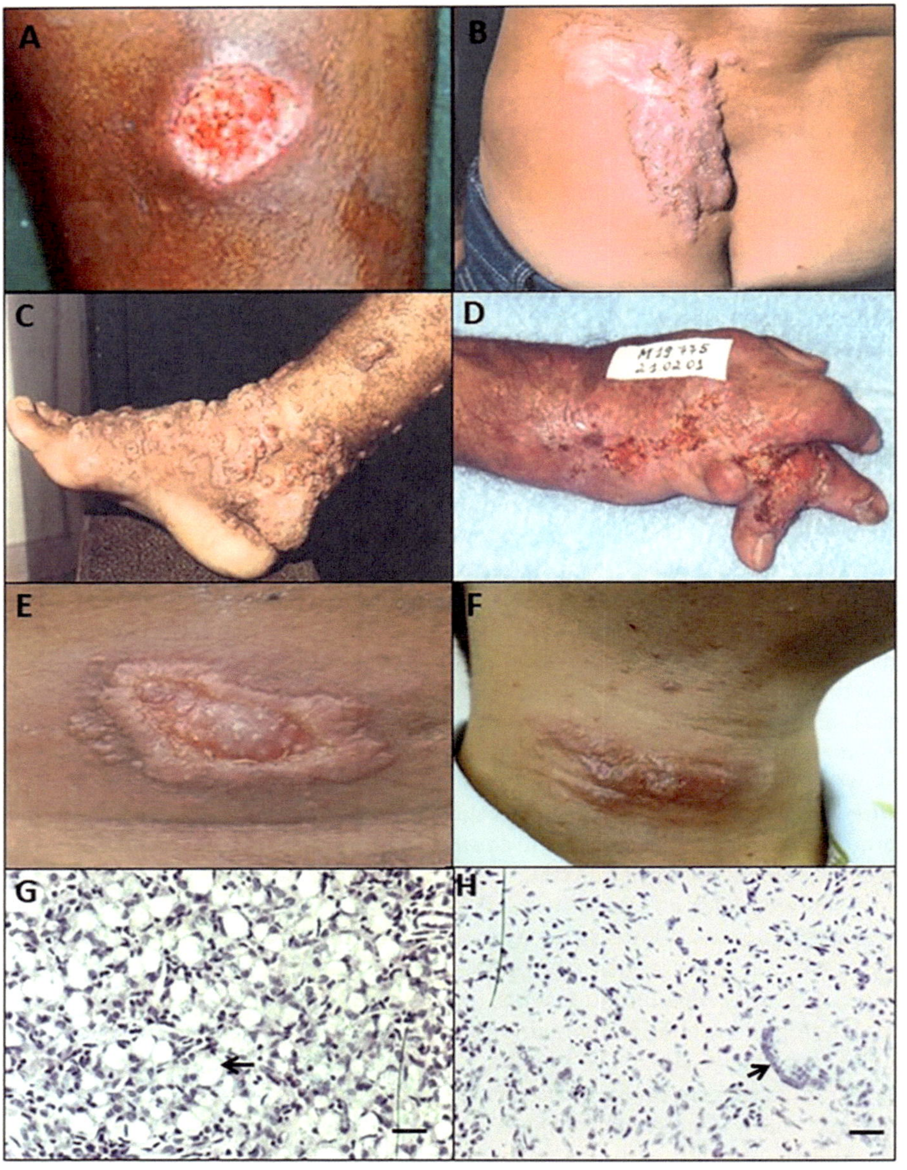

Figure 2.
Localized cutaneous leishmaniasis caused by L. (V.) braziliensis *[A: Typical ulcerated skin lesion; B: Keloid skin lesion; C: Verrucous skin lesion; D: Ulcer-infiltrated skin lesion on the hand, with bone deformity (osteomyelitis)] and* L. (L.) amazonensis *[E and F: Skin lesions in the form of a highly infiltrated plaque, with a thick base and exulcerated surface]; G and H: Histological sections of skin lesions caused by* L. (L.) amazonensis *and* L. (V.) braziliensis, *respectively, stained for eosin & hematoxylin (HE), showing in (G) dermal cellular infiltrate predominantly of heavily parasitized vacuolated macrophages (arrow), in addition to lymphocytes and plasma cells; and in (H) epithelioid granulomatous reaction, with giant cells (arrow), and lymphoplasmacytic infiltrate, with few macrophages. Bars = 20 μm.*

the epithelioid granuloma into the dermis, with the presence of giant cells, some necrosis areas, typical lymphoplasmocytic cellular infiltrate and scanty parasitized macrophages (**Figure 2H**) [59, 60]. From the immunopathological point of view, some parameters previously studied deserve to be highlighted here, such as: most patients [≥95%] present moderate to overt $DTH^{(+/++)}$, accompanied by high levels of lymphocyte proliferation assay, while into the dermis of the skin lesion the cell immunohistochemical labeling with specific monoclonal antibodies has revealed higher ($P < 0.05$) $CD8^+$ T-cells density compared to that of their $CD4^+$ T-cells counterpart, although the $CD4^+$/Th1-type immune response has been shown to outperform the $CD4^+$/Th2-type one, in addition to higher ($P < 0.05$) cell densities expressing pro-inflammatory cytokines [TNF-α, IFN-γ] compared to those of regulatory ones [IL-10, TGF-β], as well as greater $CD68^+$macrophage densities expressing *Toll-like* receptors 2 and 4 (*TLR*2, 4) compared to that of *TLR*9 were evidenced in the double immunohistochemical cell labeling (**Figure 1**) [9, 10, 18, 27, 29, 30, 53, 61–63].

On the other hand, in LCL caused by *L. (L.) amazonensis* it has also been seen that the main clinical feature of the skin lesion observed in this form of the disease is the nodule or plaque with a highly infiltrated base and edge, with a thickened appearance, evidencing discrete ulceration or even an exulceration on the surface, in which the monocyte cell infiltrate into the dermis stands out at the expense, mainly, of the non-activated vacuolated macrophage of the M2 phenotype, intensely parasitized, with few plasma cells and lymphocytes in the interstitium, characterizing a true vacuolated macrophage reaction into the dermis of patients with LCL caused by *L. (L.) amazonensis* (**Figure 2G**). It is interesting to note the harmony between these histopathological findings with those of immunopathological nature, so that, in most patients [≥80%] the delayed-type hypersensitivity reaction is absent [DTH^-] not only after challenge with axenic promastigote antigens of *L. (L.) amazonensis* or *L. (V.) braziliensis*, but also with axenic amastigote antigen of *L. (V.) lainsoni*, as well as low lymphocyte proliferation rates and few areas with epithelioid granuloma and/or necrosis into the dermis of the skin lesion in these patients are found [10, 18, 27, 51–53, 64, 65].

In addition, regarding the immunopathological findings searched by immunohistochemical cell labeling with specific monoclonal antibodies, it seems notorious certain similarity with those in LCL caused by *L. (V.) braziliensis*, such as, a higher ($P < 0.05$) $CD8^+$ T-cells density compared to that of their $CD4^+$ T-cells counterpart, as well as a $CD4^+$/Th1-type immune response over than that of $CD4^+$/Th2-type one, although not as evident as that seen in LCL caused by *L. (V.) braziliensis*. In this way, it is worth mentioning the higher ($P < 0.05$) cell densities expressing some regulatory cytokines [IL-10, TGF-β] than that of pro-inflammatory one [TNF-α], but not than that of IFN-γ. It is quite possible that the significant expression of IFN-γ together with that of *TLR*2 [evidenced by double immunohistochemical cell labeling] has ensured the superiority of the CD4+/Th1-type immune response over the CD4+/Th2-type in LCL caused by *L. (L.) amazonensis*, although some findings seem to signal a persistent suppression of the T-cell immune response as evidenced by high prevalence [≥80%] of negative $DTH^{(-)}$, as well as low rates of lymphocyte proliferation assay in these patients (**Figure 1**) [10, 18, 27, 29, 30, 32, 51, 53].

In summary, taking into account these clinical, histopathological and immunopathological features of LCL caused by *L. (V.) braziliensis* and *L. (L.) amazonensis*, it would be appropriate to say that: i) the ulcerated character of the skin lesion is more associated with *L. (V.) braziliensis*-infection [in *L. (L.) amazonensis*-infection, the skin lesion is less ulcerated, with a more infiltrated and thickened border], ii) where the presence of parasitized macrophages into the dermis is more scanty, iii) frequently [≥95% of cases] associated to positive $DTH^{(+/++)}$ [unlike *L. (L.) amazonensis*-infection with negative $DTH^{(-)}$ in ≥80% of cases] and high rates of lymphocyte proliferation assay, iv) with inflammatory response mainly represented by the epithelioid granuloma, giant cells, necrosis areas and consistent lymphoplasmacytic cell infiltrate [in *L. (L.) amazonensis*-infection, the inflammatory response is basically formed by non-activated vacuolated macrophages of the M2 phenotype, intensely parasitized, with few plasma cells and lymphocytes, characterizing a true vacuolated macrophage reaction], v) while the immunopathological scenario shows a significant protector role of $CD8^+$ T-cells stands out, in greater density than their $CD4^+$ T-cells counterpart, both in LCL by *L. (V.) braziliensis* and *L. (L.) amazonensis*, vi) the $CD4^+$/Th1-type immune response over than that of its $CD4^+$/Th2-type one, more evident in *L. (V.) braziliensis*-infection, vii) with higher cell densities of pro-inflammatory cytokines [IFN-γ, TNF-α] compared to those of regulatory ones [IL-10, TGF-β], in agreement with greater expressions of *TLR*2, 4 than that of *TLR*9 [which has shown higher interaction with *L. (L.) amazonensis*-infection, in which there has been demonstrated high prevalence [80%] of the T-cell immune response suppression].

2.2 Borderline disseminated cutaneous leishmaniasis (BDCL) due to *L. (V.) braziliensis* and *L. (L.) amazonensis*

The first point that must be clarified is why the term "borderline" is being used here to designate this form of the disease that is referred by most authors as "Disseminated cutaneous leishmaniasis" [66–73]. In fact, this term was first used by Moriearty et al. [74] to designate a clinical case of ACL in a patient with long-lasting disseminated skin lesions, and atypical intradermal reaction to *Leishmania*-antigen [DTH−] from northeast Brazil [Três Braços municipality, Bahia State], where the species *L. (V.) braziliensis* predominates. Considering that this patient also did not respond to other intradermally tested antigens, his ACL clinical form was labeled by these authors as "Borderline cutaneous leishmaniasis" [74]. However, considering that this was a typical clinical case with long-lasting disseminated skin lesions, the first step would be to designate the case as "Borderline disseminated cutaneous leishmaniasis". Anyway, considering that the term "borderline" was used by these authors in an attempt to characterize a probable patient's T-cell immune response dysfunction against infection, a similar interpretation was made by the authors of this chapter to also characterize a partial or incomplete suppression of the T-cell immune response against *Leishmania*-infection [mainly evidenced by suppression of the delayed-type hypersensitivity reaction (DTH) and lymphocyte proliferation assay] in some individuals with long-lasting disseminated skin lesions caused by *L. (V.) braziliensis* or *L. (L.) amazonensis*, favoring the spread of infection and the consequent establishment of the clinical form that has been designated by the authors of this chapter as "borderline disseminated cutaneous leishmaniasis" [BDCL] [10, 18, 27–29].

The second point that deserves to be clarified refers to the physiopathogenic mechanisms involved in the process of the parasite dissemination, which, once again, seems to be strongly influenced by the species-specific antigens of *L. (V.) braziliensis* and *L. (L.) amazonensis*. In this way, what stands out the most is the dynamics of the parasite dissemination and its relationship with the partial or incomplete suppression of the patient's T-cell immune response. In terms of BDCL caused by *L. (V.) braziliensis*, it has been observed that the parasite dissemination is relatively fast, so that in around three months, tens or even hundreds papular, acneiform or ulcerative skin lesions mainly can be counted, distributed in the vicinity of the primary skin lesion [spread by contiguity] or in different regions of the patient's body [lymphohematogenous systemic spread] (**Figure 3A–D**). Furthermore, recent evidence has shown that *L. (V.) braziliensis* isolates from patients with BDCL from an endemic area in the municipality of Três Braços, Bahia State, in northeastern Brazil, were able to induce, *in vitro*, higher infectivity rates and greater expressions of *TLR*2, 4 than isolates from LCL patients, which reinforces the central idea of this chapter on the crucial role of species-specific antigens of *L. (V.) braziliensis* and *L. (L.) amazonensis*, or even among intra-specific isolates of these *Leishmania* species in the pathogenesis of ACL. Another alternative would be that these *L. (V.) braziliensis* polymorphic variants would be producing lower systemic levels of TNF-α and IFN-γ in BDCL patients than those LCL ones, which may account for parasite dissemination due to the decreased ability to control parasite growth [18, 27, 29, 66, 68, 69, 71, 75–77].

Regarding BDCL caused by *L. (L.) amazonensis*, however, the parasite dissemination is clearly slower, taking around a year for the first metastatic lesions to appear, mainly papules that progress to non-ulcerated nodules and/or infiltrated plaques [more than ten were never observed], also close to the primary skin lesion [spread by contiguity] or at greater distances [lymphohematogenous systemic spread] (**Figure 3E–H**). Therefore, the clinical-biological behavior of *L. (L.) amazonensis*-infection, mainly with regard to the parasite dissemination and its relationship with T-cell immune response, is quite different from that by *L. (V.) braziliensis*; whereas *L. (V.) braziliensis*-infection disseminates faster and exerts a more discrete suppression of the T-cell immune response [$DTH^{\pm}$], *L. (L.) amazonensis*-infection disseminates more slowly, but with a more intense suppression of the T-cell immune response [DTH^{-}], which is reflected in the response to antimony therapy as patients with BDCL caused by *L. (L.) amazonensis* require double [four series] the dosage of antimony therapy used in the treatment of patients with BDCL caused by *L. (V.) braziliensis* [see that dosage below]. Additionally, it is interesting to note that, similarly to what has been shown in *L. (V.) braziliensis* isolates from BDCL patients, it has already been shown, also *in vitro*, greater infectivity rates of *L. (L.) amazonensis* isolates from BDCL than those from LCL, confirming our previous suspicious that intra-specific antigenic differences within *L. (L.) amazonensis* [or even within *L. (V.) braziliensis*] strains/races/populations from distinct geographical regions in Brazil may be influencing the pathogenesis of ACL [10, 18, 27–29, 38, 78].

As for the histopathological repercussion of these processes, the findings into the dermis of the skin lesions in BDCL patients do not differ much in qualitative terms from those already mentioned in LCL caused by *L. (V.) braziliensis* and *L. (L.) amazonensis*, however, it is necessary to emphasize a greater intensity of these findings that take part in these processes. For example, in BDCL caused by *L. (V.) braziliensis* what stands out is the presence of a more consistent lymphoplasmacytic cell infiltrate, accompanied by few to moderate parasitized macrophages, with few necrosis areas and epithelioid granuloma (**Figure 4B**). In BDCL caused by *L. (L.) amazonensis*, what also stands out is the exuberance of the vacuolated macrophage reaction [mainly from the M2 phenotype], intensely parasitized, accompanied by moderate lymphoplasmacytic cellular infiltrate, without necrosis areas, but in some cases with long-lasting evolution [more than a year] there may be found some epithelioid granulomas with giant cells (**Figure 4A**), which seem to represent a T-cell immune response reserve, since, as in LCL caused by this *Leishmania* species or by *L. (V.) braziliensis*, antimony therapy [12 mg/Sb^{V}/kg weight/22 days each series] has been used successfully at twice [four series] the dosage of that used in LCL [two series] [9, 10, 18, 27–29, 64].

In terms of the immunopathology of BDCL, it should be emphasized that during the parasite dissemination, which means a critical stage of the infection, both the DTH reaction and lymphocyte proliferation assay remain negative, clearly reflecting

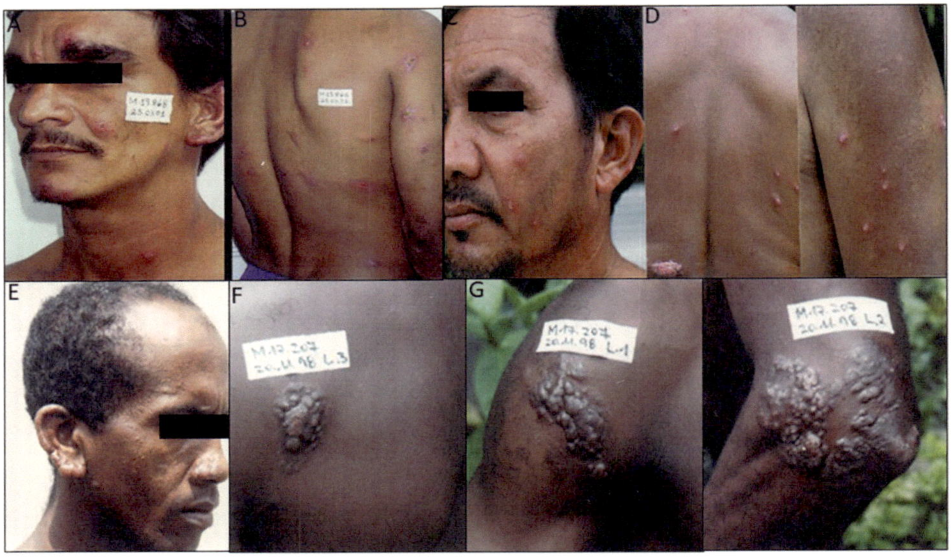

Figure 3.
Borderline disseminated cutaneous leishmaniasis (BDCL) caused by L. (V.) braziliensis *and* L. (L.) amazonensis, *respectively, with disseminated acneiform ulcerative skin lesions in A, B, C and D [L.(V.)b]; and infiltrated plaque or nodular (non-ulcerated) skin lesions in E, F, G, and H [L.(L.)a].*

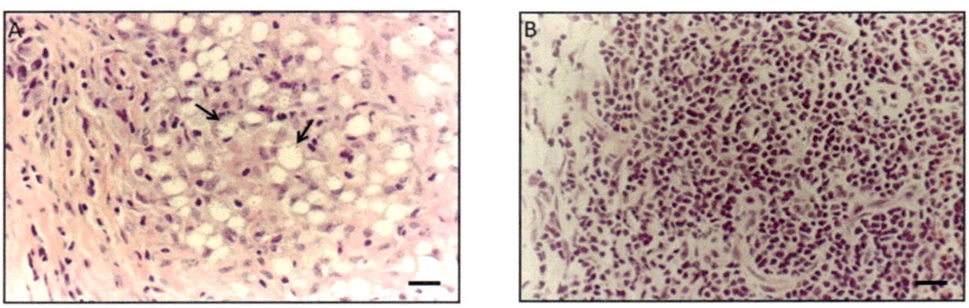

Figure 4.
Histological sections of skin lesions stained for eosin & hematoxylin (HE) in borderline disseminated cutaneous leishmaniasis (BDCL) caused by L. (L.) amazonensis *(A) and* L. (V.) braziliensis *(B), where can be seen in a typical vacuolated macrophage reaction, intensely parasitized (arrows), with few lymphocytes and plasma cells and, in B intense lymphoplasmacytic infiltrate with few macrophages. Bars = 20 μm.*

some degree of T-cell immune response suppression, mainly in those BDCL patients due to *L. (L.) amazonensis* whom present a more intense T-cell immune response suppression [DTH$^-$] than those due to *L. (V.) braziliensis* whom show a more discrete T-cell immune suppression [DTH$^{\pm}$], which, recently, was also noted by higher ($P < 0.05$) cell densities expressing regulatory cytokines [IL-10, TGF-β] than those pro-inflammatory one [TNF-α] in BDCL caused by both *L. (V.) braziliensis* and *L. (L.) amazonensis*. Moreover, it is interesting to point out that regarding the CD4$^+$/CD8$^+$ T-cells expressions in BDCL, it was also recently recorded that in *L. (V.) braziliensis*-infection there was a higher ($P < 0.05$) CD4$^+$ T-cells expression than that of their CD8$^+$ T-cells counterpart, while in *L. (L.) amazonensis*-infection there was a lower ($P < 0.05$) CD4$^+$ T-cells expression than that of their CD8$^+$ T-cells counterpart, which suggests a crucial role of CD4$^+$ T-cell immune response against infection, since in BDCL caused by *L. (L.) amazonensis*, where there was a more intense T-cell immune response suppression [DTH$^-$], the CD4$^+$ T-cells expression was lower ($P < 0.05$) than that of their CD8$^+$ T-cells one, although the CD4+ Th1-type immune response is still slightly over than the CD4+ Th2-type immune response. Finally, looking at the expressions of *TLR*2, 4 and 9, it was seen once again that *TLR*2,4 had greater ($P < 0.05$) expressions in BDCL by *L. (V.) braziliensis*, where the T-cell immune response suppression had less impact [DTH$^\pm$], while *TLR*9 had greater ($P < 0.05$) expression in BDCL by *L. (L.) amazonensis*, where T-cell immune response suppression was more impactful [DTH$^-$] (**Figure 1**) [9, 10, 18, 27–30].

In summary, BDCL represents an intermediary form of ACL that can occupy the two places between the center (LCL) and the two polar forms (MCL/ML and ADCL) in the clinical-immunopathological spectrum of the disease, having *L. (V.) braziliensis* and *L. (L.) amazonensis* as the main etiological agents in Brazil, as well as there are some clinical and immunopathological features that differ BDCL between these two *Leishmania* species. The term "borderline" was introduced to characterize the clinical evidence of partial or incomplete T-cell immune response suppression against *Leishmania*-infection in BDCL, mainly in those patients infected by *L. (L.) amazonensis* who present greater suppression of T-cell immune response [DTH$^-$] than those infected by *L. (V.) braziliensis* [DTH$^\pm$], which seems to be one of the main reasons for the spread of the parasite. However, in BDCL by *L. (V.) braziliensis*, the parasite spread is relatively rapid, requiring only three months to give rise to a hundred or more ulcerative skin lesions; in contrast, in BDCL by *L. (L.) amazonensis*, the parasite spread is slower, requiring around one year to produce some 5 to 10 infiltrated skin lesions. As for the histopathology, attention is drawn to the exuberance of the vacuolated macrophage reaction, mainly at the expense of the intensely parasitized M2 phenotype macrophage in the cellular infiltrate of BDCL by *L. (L.) amazonensis*, while in BDCL by *L. (V.) braziliensis* what stands out is the presence of the characteristic lymphoplasmacytic cell infiltrate with few parasitized macrophages. These histopathological findings seem to be in agreement with those immunopathological ones that revealed higher [$P < 0.05$] expression of regulatory [IL-10 and TGF-β] than pro-inflammatory [TNF-α] cytokines in BDCL by both *L. (V.) braziliensis* and *L. (L.) amazonensis*, but with higher [$P < 0.05$] expression of CD4$^+$ T-cells than their CD8$^+$ T-cells counterpart in BDCL caused by *L. (V.) braziliensis*, although the opposite was observed in BDCL caused by *L. (L.) amazonensis* [lower expression of CD4$^+$ T-cells than their CD8$^+$ T-cells counterpart], which reinforces a higher T-cell immune response suppression in BDCL by *L. (L.) amazonensis*. However, in spite of these observations, the partial or incomplete T-cell immune response suppression found in BDCL by *L. (L.) amazonensis* has been restored following successful treatment with four LCL-therapeutic antimony regimens [12 mg/Sb^V/kg weight/22 days each series with a 10-day interval between series], which suggest an evident supremacy of Th1 over Th2 immune response (Th1 $\geq$ Th2).

2.3 Mucocutaneous or mucosal leishmaniasis (MCL/ML) due to *L. (V.) braziliensis*

The first point that needs to be clarified concerns the possible similarities or differences between the both terms mucocutaneous and mucosal leishmaniasis. What is common to both is the involvement of nasobuccopharyngeal mucosal tissue, or just nasal [which is more frequent] or just buccopharyngeal, however, what distinguishes them is the dynamics of infection, since in some situations the infection may present simultaneous involvement of the skin and nasobuccopharyngeal mucosa tissue [mucocutaneous leishmaniasis - MCL] or only the nasobuccopharyngeal mucosa tissue [mucosal leishmaniasis - ML]. Thus, it seems clear that in the case of MCL, the involvement of the mucosal tissue is earlier (a few months after the appearance of the skin lesion) than in ML, in which the appearance of the mucosal lesion may be after years (average two to three years) of apparent healing (spontaneous or after irregular therapy) of the skin lesion. Nevertheless, it is necessary to say that in the Brazilian Amazon, especially in the State of Pará, the form more frequently observed is the ML, while the MCL form is more prevalent in the Brazilian northeast, especially in the State of Bahia, which suggests, once again, that intra-specific antigenic differences within *L. (V.) braziliensis* strains/races/populations from distinct geographical regions in Brazil may be influencing the pathogenesis of ACL [MCL/ML] [1, 2, 10, 18, 27, 43, 79–83].

Anyway, what needs to be emphasized is the extraordinary capacity of *L. (V.) braziliensis* to migrate, via lymphohematogenous, from a skin lesion to the nasobuccopharyngeal mucosa tissue, where it will develop simultaneously to the skin lesion(s) [in the case of MCL], or years, in general, after apparent spontaneous remission or the inadequate treatment-based healing of skin lesion [in the case of ML], an inflammatory process that begins with erythema and infiltration of the mucosal tissue, followed by granulation and ulceration of the cartilaginous tissue, giving an ulcerogranulomatous aspect to the mucosal lesion (**Figure 5A–C**). In advanced cases, usually with more than five years of evolution, the inflammatory process can promote uncomfortable conditions such as nasal obstruction, formation of crusts, epistaxis, rhinorrhea, pruritus and even perforation of the cartilaginous tissue of the nasal septum or palate. In order of frequency, mucosal lesions primarily occur in the nasal septum, palate, pharynx, and larynx, or may be simultaneous in some cases resulting in metastatic lesions of the nasobuccopharyngeal mucosal tissue [1, 2, 4, 6, 10, 18, 27, 43, 83, 84].

Regarding the histopathology of MCL/ML, the most striking finding of this form of the disease is represented by the presence of the tuberculoid granuloma in the nasobuccopharyngeal mucosal tissue, which corroborates not only the remarkable $DTH^{(++++)}$ and lymphocyte proliferation assay, but also the up-regulation of $CD4^+$/Th1-type immune response observed in MCL/ML. In addition, other histopathological findings include abundant infiltration of lymphocytes and plasma cells, with scanty parasitized macrophages, followed by necrosis of the cartilaginous structures (**Figure 5D**) [31, 53, 59, 60].

In parallel with the histopathology of nasobuccopharyngeal mucosal lesions, it is important to note that some immunohistochemical studies have shown that MCL/ML is supported not only by the increased expression of $CD4^+$ T-cells infiltrating the mucosal lesions of patients (as compared to corresponding $CD8^+$ T-cells), as well as the increased expression of pro-inflammatory TNF-α and IFN-γ as compared to regulatory TGF-β and IL-10 cytokines and lysozyme stained activated macrophages (epithelioid cells), characterizing a typical $CD4^+$/Th1-type immune response, which reinforces a prior hypothesis regarding the pathogenesis of MCL/ML caused by *L. (V.) braziliensis* that implies the activation of a persistent antigen-specific $CD4^+$/Th1-type immune response at the lymph nodes, followed by the recruitment of $CD4^+$/T-cells from the peripheral blood to the inflammatory infiltrate in the mucosal lesions. As such, those cells are preferentially primed to operate as cytokine Th1-producing cells (mainly IFN-γ and TNF-α). The strong T-cell hypersensitive response recorded in these patients can

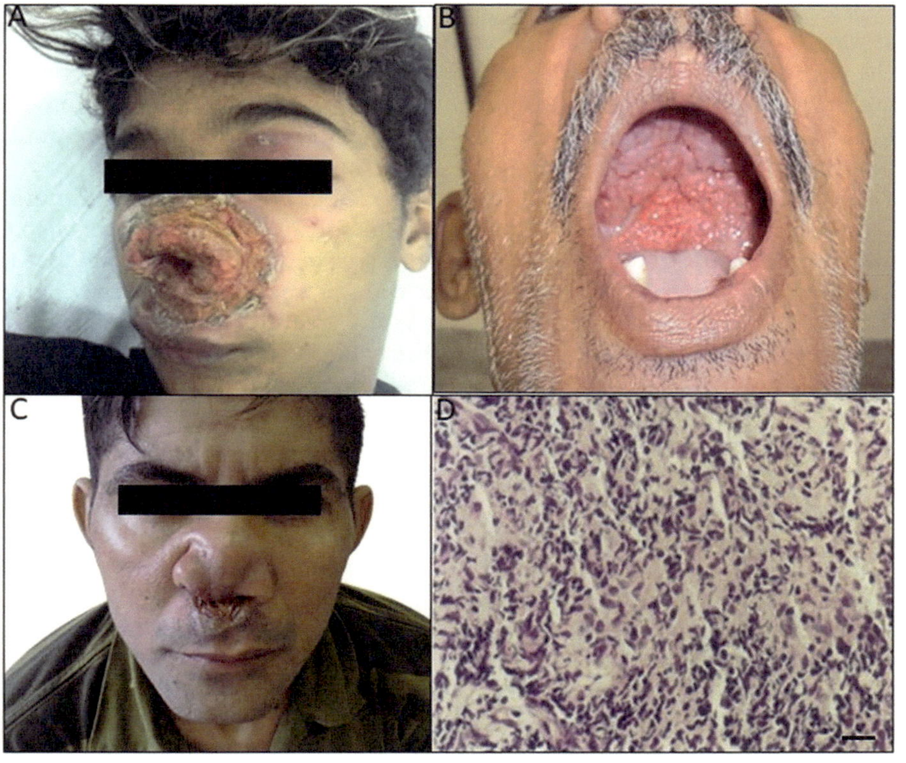

Figure 5.
Clinical aspects of mucocutaneous (MCL) or mucosal leishmaniasis (ML) caused by L. (V.) braziliensis *in A, B and C, and in D a histological section of a mucosal lesion stained for eosin & hematoxylin [EH] showing a garnulomatous reaction with intense lymphoplasmacytic infiltrate. Bar = 20 μm.*

be attributed to a long-lasting antigen-specific $CD4^+$/Th1 activation by species-specific *L. (V.) braziliensis*-antigens, resulting in the high production of IFN-γ and TNF-α cytokines. In addition, the moderate expression of *TLR*2, 4 should also be highlighted, however, higher than that of *TLR*9 in MCL, which reinforces the greater interaction of *TLR*2,4 with *L. (V.) braziliensis*-infection than *TLR*9 (**Figure 1**) [10, 18, 27, 29–31, 85]. TNF-α seems to be the major cytokine responsible for damaging the mucosal tissue [86–89]. Corroborating these findings, it should also be noted recent findings on the gene expression of immune response in ACL by *L. (V.) braziliensis*, showing high levels of IFN-γ and TNF-α cytokines in older skin lesions containing detectable levels of parasite transcripts [90].

2.4 Anergic diffuse cutaneous leishmaniasis (ADCL) due to *L. (L.) amazonensis*

Anergic diffuse cutaneous leishmaniasis (ADCL) is regarded a rare clinical form of ACL caused by leishmanine parasites belonging to the subgenera *Leishmania* and *Mundinia*. In this way, although *L. (L.) mexicana* and *L. (L.) amazonensis* have been regarded as the major agents of ADCL in Latin America, *L. (M.) martiniquensis* and *L. (L.) waltoni* have also been incriminated, more recently, as etiologic agents of the disease in that region [12, 14, 44, 91–93].

In Brazil, principally in the Brazilian Amazon, *L. (L.) amazonensis* is regarded, so far, the only species with immunopathogenic competence to determine ADCL; that is, with the ability to deviate the T-cell immune response of the infected individual to the T-cell hyposensitivity pole [DTH−] of the clinical-immunopathological ACL spectrum, characterized by strong expression of the T-cell immune response of $CD4^+$/Th2-type, with high production of well-known regulatory cytokines [IL-10 and TGF-β]. It is in this subverted or down-regulated immunopathological scenario, strongly influenced by the species-specific action of *L. (L.) amazonensis*-antigens, that the infection progresses to establish a clinical picture characterized by a diffuse infiltration of the skin, as well as by papular, tuberous, infiltrating and nodular skin lesions, which rarely ulcerate, distributed throughout the integument only sparing the scalp and the palmar (hand) and plantar (foot) regions (**Figure 6A–C**). Due to this persistent immunosuppressive character of $CD4^+$/Th2-type immune response, most patients (mainly children under five years old) carry, for many years (sometimes four or more decades), a severe clinical picture, sometimes highly mutilating, leading to the chronification of the process since the drugs available (pentavalent antimony, amphotericin B, pentamidine and miltefosine) for the treatment are not able to change the immunogenetic status of the patients. In cases with advanced evolution (over thirty years), bone lesions may occur in the extremities (osteomyelitis) or involvement of the nasobuccopharyngeal mucosa due to the parasite dissemination by contiguity of lesions on the face [10, 18, 27, 29, 30, 43, 94–99].

Based on the immunosuppressive character of the disease, it would not be difficult to expect that the major histopathological finding of the skin lesions would reveal a monocytic cellular infiltrate into the dermis predominantly composed by vacuolated macrophages with a M2 phenotype, intensely parasitized, in addition to a moderate presence of plasma cells and few lymphocytes (**Figure 6D**) [10, 18, 27, 28, 43, 52, 100, 101]. In general, this is the quasi-stigmatic histopathological picture of ADCL that together with the persistent suppressive $CD4^+$/Th2-type immune response explain why the authors of this chapter use the term "anergic" to name this type of ACL with a prognosis that is still quite reserved.

From an immunopathological point of view, there is no doubt that the specific suppression of the T-cell immune response [$CD4^+$/$CD8^+$] provided by the species-specific *L. (L.) amazonensis*-antigens has wide repercussions on different fronts of the immune defense mechanisms against infection, evidenced by the strong suppression of the delayed-type hypersensitivity reaction (DTH^-) and lymphocyte proliferation responses, as well as by the low production, *in situ*, of the main inflammatory cytokines [IFN-γ and TNF-α] and, in contrast, the high production of regulatory ones [IL-4, IL-10 and TGF-β], revealing the inability of the non-activated dermal macrophage, mainly the M2 phenotype, to contain the parasite multiplication and leading to very weak responses to conventional therapy regimens [antimony, pentamidine and amphotericin B]. Corroborating these findings, it is also worth mentioning that it is precisely in ADCL that the lowest expression of $CD4^+$/$CD8^+$ T-cells is shown among all ACL clinical forms, but, on the contrary, it is also in ADCL that the highest expression of Treg $Foxp3^+$ regulatory cells is recorded, confirming the extremely suppressed profile of $CD4^+$/Th2-type immune response in ADCL, which seems to be strongly associated to low transcripts of $CD4^+$/Th1-type and cytotoxic $CD8^{+/}$T-cells and the highest expression of *TLR9* in the clinical-immunopatholgical ACL spectrum [18, 27, 29, 30, 38, 51, 53, 64, 102, 103].

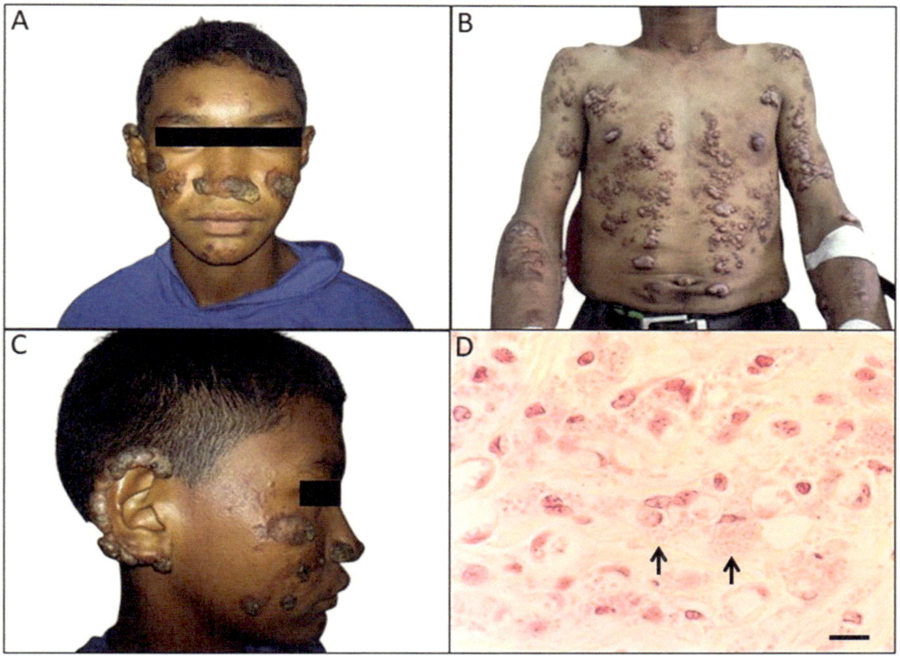

Figure 6.
Clinical aspects of anergic diffuse cutaneous leishmaniasis (ADCL) caused by L. (L.) amazonensis *in A, B and C, and in D a histological section of a skin lesion stained for eosin & hematoxylin (HE) exhibiting diffuse vacuolated macrophage reaction, intensely parasitized (arrows), with few lymphocytes and plasma cells. Bar = 20 μm.*

3. Importance of species-specific antigens of *L. (V.) braziliensis* and *L. (L.) amazonensis* in the pathogenesis of ACL

Leishmania parasites are well-known as obligate intracellular protozoan organisms because they carry out an essential phase of their reproduction within the mononuclear phagocytic cells, mainly the macrophage, in which the amastigote form of the parasite multiplies by binary division [14]. However, in order for the amastigote form of the parasite to resist the lethal action of a highly hostile environment inside the macrophage's phagolysosome [highly acidic, oxidative, containing antimicrobial peptides, proteases and restrict nutrient access], the parasite makes use of a series of molecules on the surface of its plasma membrane, called glycoconjugates, in order to neutralize the lethal action of these hostile elements into the macrophage's phagolysosome [104, 105]. This is a critical phase for the parasite that will result in its survival or death, so that, if overcome by the parasite, the infection will be definitively established in the vertebrate host [106, 107]. However, in addition to the protective effect that these glycoconjugate molecules exert on the survival of the parasite, it is also known that their action can play an important role in the virulence of the parasite, reason why these glycoconjugate molecules are recognized here as species-specific *Leishmania*-antigens of both subgenera *Leishmania* and *Viannia*, with pivotal role in the pathogenesis of the disease caused by these *Leishmania* parasites, represented here by *L. (V.) braziliensis* and *L. (L.) amazonensis*.

Among the previously mentioned glycoconjugate molecules [LPG, GIPLs, PPGs, GP-63, PS and CD200], LPG is undoubtedly one of most important species-specific *Leishmania*-antigens or, in other words, the major cell surface glycoconjugate of *Leishmania* parasites, which has the ability to protect the parasite against complement-mediated lysis, facilitating its attachment and entry into the macrophage, and inhibiting the phagolysosomal fusion and protein kinase C [108–110]. Furthermore, it seems to exert one of the broadest antigenic activity spectrum being able to induce not only a pro-inflammatory innate immune response, but also a suppressive one, depending on the subgenus of *Leishmania* species involved in the process, i.e., *Leishmania* or *Viannia*. In this sense, one cannot neglect the possibility that the species-specific antigens of these subgenera (*Leishmania* and *Viannia*) may be true messengers of glycoconjugate molecules formed from the contents of the sandfly's digestive tract segment where the parasite reproduced; that is, that of the subgenus *Leishmania* in the midgut [suprapylaria], while that of the sugenus *Viannia* in the posterior intestine [peripylaria] [14]. It is possible that this may be influencing the immunobiological role of the glycoconjugate molecules that act as species-specific antigens for *Leishmania* species of the *Leishmania* (*L. amazonensis*) and *Viannia* (*L. braziliensis*) subgenera.

In this way, it has been shown that LPG exerts a pivotal role in modulating a cascade of activated inflammatory T-cell responses in experimental *L. (V.) braziliensis*-infections, including high expression of TNF-α, IL-1β, IL-6 and NO (such as those seen in ML). In addition, it should also been highlighted that *L. (V.) braziliensis*-LPG not only stimulates messenger RNA transcription, but also induces NF-kB translocation, resulting in NO expression as well as IFN-γ and TNF-α cytokines induction in ACL. That induction is mediated by linking *L. (V.) braziliensis*-LPG to *TLR*2 or *TLR*4, indicating the crucial role of those receptors in innate defenses, which was also evidenced in clinical samples of LCL caused by *L. (V.) braziliensis* [30, 34, 36]. On the other hand, LPG has also been shown to act as an important immunomodulatory factor activating *TLR*4 and promoting the preferential $CD4^{+}$/Th2-type immune response in experimental *L. (L.) amazonensis*-infections, which is strongly related to ADCL [35]. Based on these results, the most emphatic impression left is that the PLG molecule seems to act in a bidirectional manner from an immunobiological point of view, depending on the subgenus, *Leishmania* [*L. amazonensis*] or *Viannia* [*L. braziliensis*], involved in the process, although, more recently, it has been shown that LPG of *L. (L.) amazonensis* isolates from different origins, that is, from an ADCL case (State of Bahia, Brazil), from a domestic dog with visceral leishmaniasis (State of Minas Gerais, Brazil) and from a wild rodent (*Proechmys* sp.) in the Brazilian Amazon, showed a significant pro-inflammatory response with expression of TNF-α and IL-6 cytokines [111].

Another interesting species-specific *Leishmania*-antigen is the PS molecule, a phospholipid commonly located on the inner face of the parasite plasma membrane with capacity for inhibiting the inflammatory response of *Leishmania*-infected macrophages (as a strategy to evade the host T-cell immune response). In that sense, higher PS exposure on the surface of late-stationary-phase promastigotes of *L. (V.) braziliensis* has been observed in patients with LCL as compared to patients with ML [112]. However, higher PS exposure has been noted in *L. (L.) amazonensis* isolates obtained from ADCL patients as compared to those from LCL patients [113]. Those findings represent consistent evidence that the variability of PS across different strains of *Leishmania* [*L. amazonensis*] and *Viannia* [*L. braziliensis*] subgenera can influence the development of different clinical forms of ACL.

Finally, it is worth mentioning the important role that the CD200 molecule plays in the pathogenesis of *L. (L.) amazonensis*-infection, which is capable of inhibiting the activation of the infected macrophage via interaction with *TLR*9; that is, the CD200-dependent iNOS inhibition allows parasite growth within the macrophage, increasing the virulence of *L. (L.) amazonensis* [114]. Those findings were recently confirmed clinically, with evident interactions of *TLR*2 with clinical forms associated with *L. (V.) braziliensis*, including LCL, BDCL, and ML, while *TLR*9 showed strong interactions with clinical forms associated with *L. (L.) amazonensis* (principally the down regulated ADCL form) [30].

4. Concluding remarks

Based on what was mentioned above, there seems to be no doubt about how important it is to know the biology of the parasites of the genus *Leishmania* in order to better understand their relationships not only with their primitive hosts, invertebrates [phlebotomine sand flies] and vertebrates [wild mammals], but also with man, who is merely an accidental host [14]. This knowledge today has gained a great advance, mainly helped by the development of molecular biology, which has allowed to deepen the knowledge of the biology of these parasites at the level of the molecules of their plasma membrane' (glycoconjugates), allowing a better understanding of their more specific relationships with the human immune response, mainly the T-cell immune responses, CD4/Th1-type and CD4/Th2-type, whose modulation by these glycoconjugate molecules, referred to herein as species-specific antigens of *Leishmania* species of the *Leishmania* [*L. amazonensis*] and *Viannia* [*L. braziliensis*] subgenera, will be of the same greater importance regarding the modulation of the clinical-immunopathological spectrum of ACL in Brazil [10, 18, 27, 29, 30].

Acknowledgements

This work was supported by Evandro Chagas Institute (Secretary of Health and Environment Surveillance of Ministry of Health, Brazil); Tropical Medicine Nucleus (Federal University of Pará State, Brazil); and "Laboratório de Patologia de Moléstias Infecciosas (LIM-50), Hospital das Clínicas HCFMUSP, Faculdade de Medicina, Universidade de Sao Paulo, SP, BR, and Grant #2014/50315-0, São Paulo Research Foundation (FAPESP)".

Conflict of interests

The author declares no conflict of interest.

Author details

Fernando T. Silveira[1,2]*, Marliane B. Campos[1], Silvia F. Müller[3], Patrícia K. Ramos[1], Luciana V. Lima[1], Thiago V. dos Santos[1], Claudia Maria Gomes[4], Márcia D. Laurenti[4], Vania Lucia da Matta[4] and Carlos Eduardo Corbett[4]

1 Parasitology Department, Evandro Chagas Institute (Secretary of Science, Technology and Innovation, Ministry of Health), Ananindeua, Pará State, Brazil

2 Tropical Medicine Institute, Federal University of Pará, Belém, Pará State, Brazil

3 Dermatology Service, Federal University of Pará, Belém, Pará State, Brazil

4 Pathology Laboratory of Infectious Diseases (LIM50), Pathology Department, Medical School of São Paulo University, São Paulo, São Paulo State, Brazil

*Address all correspondence to: fernandotobias@iec.gov.br

References

[1] Marsden PD. Mucosal leishmaniasis ("espundia" Escomel, 1911). Transactions of the Royal Society of Tropical Medicine and Hygiene. 1986;**80**:859-876

[2] Marsden PD. Mucocutaneous leishmaniasis. British Medical Journal. 1990;**301**:656-657

[3] Marsden PD, Llanos-Cuentas EA, Lago EL, et al. Human mucocutaneous leishmaniasis in Três Braços, Bahia-Brazil. An area of *Leishmania braziliensis braziliensis* transmission. III. Mucosal disease presentation and initial evolution. Revista da Sociedade Brasileira de Medicina Tropical. 1984;**17**:179-186

[4] Talhari S, Talhari AC, Ferreira LC, et al. Leishmaniose Cutaneomucosa. In: Talhari S, Neves RG, editors. Rio de Janeiro (RJ), Brasil: Dermatologia Tropical. Ed. Medsi; 1995. pp. 23-45

[5] Pearson RD, Sousa AQ. Clinical Spectrum of Leishmaniasis. Clinical Infectious Diseases. 1996;**22**(1):1-13. DOI: 10.1093/clinids/22.1.1

[6] Müller SFR. Atividade dos antígenos de *Leishmania* A2 e PH8 *in vitro*, na indução da resposta imunocelular. Tese (Doutorado em Pós Graduação Área de Clínica Médica). São Paulo: Faculdade de Medicina de Ribeirão Preto, Universidade de São Paulo (USP); 2004. p. 119 f

[7] Amato VS, Tuon FF, Bacha HA, Neto VA, Nicodemo AC. Mucosal Leishmaniasis. Current scenario and prospects for treatment. Acta Tropica. 2008;**105**(1):1-9. DOI: 10.1016/j.actatropica.2007.08.003

[8] Handler MZ, Patel PA, Kapila R, Al-Qubati Y, Schwartz RA. Cutaneous and Mucocutaneous Leishmaniasis: Clinical perspectives. Journal of the American Academy of Dermatology. 2015;**73**(6):897-908; quiz 909-10. DOI: 10.1016/j.jaad.2014.08.051

[9] Silveira FT, Lainson R, Muller SFR, et al. Leishmaniose tegumentar Americana. In: Leão RNG, editor. Medicina Tropical e Infectologia na Amazônia. 1a ed. Vol. 2. Belém, Pará, Brasil: Instituto Evandro Chagas; 2013. pp. 1203-1244

[10] Silveira FT, Muller SFR, Laurenti MD, et al. Leishmaniose Tegumentar Americana. In: Tratado de Dermatología. Belda Junior W, Di Chiacchio N, Criado PR (Ed). 3ª Ed. Cap. vol. 72. São Paulo: Atheneu; 2018. pp. 1691-1700

[11] Lainson R. The Neotropical *Leishmania* species: A brief historical review of their discovery, ecology and taxonomy. Rev Pan-Amaz Saúde. 2010;**1**:13-38

[12] Lainson R, Shaw JJ. New World Leishmaniasis. In: Collier L, Balows A, Sussman M, editors. Topley and Wilson's Microbiology and Microbial Infections. 10th ed. Vol. 5. London: Parasitol, Arnold; 2010. pp. 313-349

[13] Vianna G. Sobre uma nova espécie de *Leishmania* (nota preliminar). Brasil Médico. 1911;**25**:411-1911

[14] Lainson R, Shaw JJ. Evolution, classification and geographical distribution. In: Peters W, Killick-Kendrick, editors. The Leishmaniases in Biology and Medicine. London: Academic Press; 1987. pp. 1-120

[15] Lainson R. The American leishmaniasis: Some observation on their

ecology and epidemiology. Transactions of the Royal Society of Tropical Medicine and Hygiene. 1983;77:569-596

[16] Lainson R, Shaw JJ. Epidemiology and ecology of leishmaniasis in Latin America. Nature. 1978;**273**:596-600

[17] Jennings YL, de Souza AA, Ishikawa EA, et al. Phenotypic characterization of *Leishmania* spp. causing cutaneous leishmaniasis in the lower Amazon region, western Pará state, Brazil, reveals a putative hybrid parasite, *Leishmania (Viannia) guyanensis* x *Leishmania (Viannia) shawi shawi*. Parasite. 2014;**21**(39):1-11

[18] Silveira FT. What makes mucosal and anergic diffuse cutaneous leishmaniases so clinically and immunopathologically different? Transactions of Royal Society of Tropical Medical and Hygiene. 2019;**113**:505-516

[19] Espinosa OA, Serrano MG, Camargo EP, et al. An appraisal of the taxonomy and nomenclature of trypanosomatids presently classified as *Leishmania* and *Endotrypanum*. Parasitology. 2016;**15**:1-13. DOI: 10.1017/S0031182016002092

[20] Lainson R, Shaw JJ. The role of animals in the epidemiology of south American leishmaniasis. In: Biology of the Kinetoplastida. WHR, Lumsden & DA Evans Eds. Vol. 2. London, New York: Academic Press; 1979. pp. 1-116

[21] Arias JR, Freitas RAD. On the vectors of cutaneous leishmaniasis in the Central Amazon of Brazil. 3. Phlebotomine sand fly stratification in a terra firme forest. Acta Amazonica. 1982;**12**:599-608

[22] Lainson R, Shaw JJ, Silveira FT, et al. The dermal leishmaniasis of Brazil, with special reference to the eco-epidemiology of the disease in Amazonia. Memórias do Instituto Oswaldo Cruz. 1994;**89**:435-443

[23] Vasconcelos dos Santos T, Prévot G, Ginouvès M, Duarte G, Silveira FT, Póvoa MM, et al. Ecological aspects of Phlebotomines (Diptera: Psychodidae) and the transmission of American cutaneous leishmaniasis agents in an Amazonian/Guianan bordering area. Parasite Vectors. 2018;**11**:1-13

[24] Brilhante AF, Lima L, de Ávila MM, Medeiros-Sousa AR, de Souza JF, Dos Santos NP, et al. Remarkable diversity, new records and *Leishmania* detection in the sand fly fauna of an area of high endemicity for cutaneous leishmaniasis in acre state, Brazilian Amazonian Forest. Acta Tropica. 2021;**223**:106103. DOI: 10.1016/j.actatropica.2021.106103

[25] Brasil. Ministério da Saude (MS), Secretaria de Vigilância da Saúde (SVS), Dep. De Vigilância das Doenças Transmissíveis (DVDT). Manual de vigilância da leishmaniose tegumentar. Brasília: SVS, DVDT; 2017. p. 189

[26] Jones TC, Johnson WD, Barretto AC, Lago E, Badaro R, Cerf B, et al. Epidemiology of American cutaneous Leishmaniasis due to Leishmania Braziliensis Braziliensis. The Journal of Infectious Diseases. 1987;**156**(1):73-83. DOI: 10.1093/infdis/156.1.73

[27] Silveira FT, Lainson R, Corbert CEP. Clinical and immunopathological spectrum of American cutaneous leishmaniasis with special reference to the disease in the Amazonian Brazil. Memórias do Instituto Oswaldo Cruz. 2004;**99**:239-251

[28] Silveira FT, Lainson R, Corbett CEP. Further observations on clinical, histopathological and immunological features of borderline disseminated

cutaneous leishmaniasis caused by *Leishmania (Leishmania) amazonensis*. Memórias do Instituto Oswaldo Cruz. 2005;**100**:525-534

[29] Silveira FT, Lainson R, Gomes CM, et al. Immunopathogenic competences of *Leishmania (V.) braziliensis* and *L. (L.) amazonensis* in American cutaneous leishmaniasis. Parasite Immunology. 2009;**31**(8):423-431

[30] Campos MB, Lima LVR, de Lima ACS, et al. *Toll-like* receptors 2, 4, and 9 expressions over the entire clinical and immunopathological spectrum of American cutaneous leishmaniasis due to *Leishmania (V.) braziliensis* and *Leishmania (L.) amazonensis*. PLoS One. 2018;**13**:e0194383. DOI: 10.1371/journal.pone.0194383

[31] Bacellar O, Lessa H, Schriefer A, et al. Up-regulation of Th1-type responses in mucosal leishmaniasis patients. Interferon Immunity. 2002;**70**:6734-6740

[32] Silveira FT, Müller SFR, de Souza AAA, Lainson R, Gomes CMC, Laurent MD, et al. Revisão sobre a patogenia da leishmaniose tegumentar americana na Amazônia, com ênfase à doença causada por *Leishmania (V.) braziliensis* e *Leishmania (L.) amazonensis*. Revista Paraense de Medicina. 2008;**22**(1):9-20

[33] Carvalho AK, Silveira FT, Passero LFD, et al. *Leishmania (V.) braziliensis* and *L. (L.) amazonensis* promote differential expression of dendritic cells and cellular immune response in murine model. Parasite Immunology. 2012;**34**:395-403

[34] Ibraim IC, de Assis RR, Pessoa NL, et al. Two biochemically distinct lipophosphoglycans from *Leishmania braziliensis* and *Leishmania infantum* trigger different innate immune responses in murine macrophages. Parasite Vectors. 2013;**6**:1-11

[35] Nogueira PM, Assis RR, Melo MN, Saraiva EM, Pessoa NL, Campos MAS, et al. Lipophosphoglycans from *Leishmania amazonensis* strains display immunomodulatory properties via *TLR*4 and do not affect sand fly infection. PLoS Neglected Tropical Diseases. 2016;**10**(8):e0004848

[36] Vieira TS, Rugani JN, Nogueira PM, Torrecilhas AC, Gontijo CM, Descoteaux A, et al. Intraspecies polymorphisms in the Lipophosphoglycan of *L. braziliensis* differentially modulate macrophage activation via *TLR*4. Frontiers in Cellular and Infection Microbiology. 2019;**9**:240

[37] PAHO. Leishmanioses: Informe epidemiológico das Américas. Washington, D.C.: OPAS-WHO; 2019

[38] Silveira FT. Leishmaniose cutânea difusa na Amazônia, Brasil: Aspectos clínicos e epidemiológicos. Gaz Méd Bahia. 2009;**79**(Suppl. 3):25-29

[39] Xavier MB, Silveira FT, Demachki S, et al. American tegumentary leishmaniasis: A quantitative analysis of Langerhans cells presents important differences between *L. (L.) amazonensis* and *Viannia* subgenus. Acta Tropica. 2005;**95**:67-73

[40] Silveira FT, Lainson R, Gomes CMC, et al. Reviewing the role of dendritic Langerhans cells in the immunopathogenesis of American cutaneous leishmaniasis. Transactions of the Royal Society of Tropical Medicine and Hygiene. 2008;**102**:1074-1080

[41] Soong L, Henard CA, Melby PC. Immunopathogenesis of non-healing American cutaneous leishmaniasis and progressive visceral leishmaniasis.

Seminars in Immunopathology. 2012;**34**:735-751. DOI: 10.1007/s00281-012-0350-8

[42] Guerra JA, Prestes SR, Silveira H, et al. Mucosal Leishmaniasis caused by *Leishmania (Viannia) braziliensis* and *Leishmania (Viannia) guyanensis* in the Brazilian Amazon. PLos Neglected Tropical Diseases. 2011;**5**(3):e980. DOI: 10.1371/journal.pntd.0000980

[43] Silveira FT, Duarte ERL, De-Farias ECF, Ikeda CS, Lopes AP, Chagas EJP, et al. Leishmaniose mucosa na Amazônia brasileira: Avaliação, retrospectiva, dos aspectos clínicos e epidemiológicos da doença, com ênfase ao estado do Pará. Revista da Sociedade Brasileira de Medicina Tropical. 1999;**32**(Supl. I):9

[44] Lainson R, Shaw JJ. Leishmaniasis in the New World. In: Collier L, Balows A, Sussman M, editors. Topley & Wilson's *Microbiology and Microbial Infections*. 10th ed. Vol. 5. London: Arnold; 2005. pp. 313-349. Parasitology

[45] Lainson R, Shaw JJ. Leishmaniasis in Brazil. I. Observations on enzootic rodent leishmaniasis – Incrimination of *Lutzomyia flaviscutellata* (Mangabeira) as the vector in the lower Amazonian basin. Transactions of the Royal Society of Tropical Medicine and Hygiene. 1968;**62**:385-395

[46] Souza AAA, Ishikawa EAY, Braga RR, Silveira FT, Lainson R, Shaw JJ. *Psychodopygus complexus,* a new vector of *Leishmania braziliensis* to humans in Pará state, Brazil. Transactions of the Royal Society of Tropical Medicine and Hygiene. 1996;**90**:112-113

[47] Rangel EF, Lainson R. Proven and putative vectors of American cutaneous leishmaniasis in Brazil: Aspects of their biology and vectorial competence. Memórias do Instituto Oswaldo Cruz. 2009;**104**(7):937-954

[48] Souza AAA, Silveira FT, Lainson R, Barata IR, Silva MGS, Lima JAN, et al. Fauna flebotomínica da Serra dos Carajás, Estado do Pará, Brasil, e sua possível implicação na transmissão da leishmaniose tegumentar americana. Rev Pan-Amaz Saude. 2010;**1**(1):45-51

[49] Andrade ZA, Reed SG, Roters SB, Sadigursky N. Immunopathology of experimental cutaneous leishmaniasis. The Ameican Journal of Pathology. 1984;**114**:137-148

[50] Andrade ZA, Reed SG, Roters SB, Sadigursky N. Patogenia da Leishmaniose cutânea experimental. A importância da necrose na eliminação dos parasitos da lesão. Revista da Sociedade Brasileira de Medicina Tropical. 1984;**17**:187-197

[51] Silveira FT, Lainson R, Shaw JJ, et al. Cutaneous leishmaniasis due to *Leishmania (Leishmania) amazonensis* in Amazonian Brazil, and the significance of a Montenegro skin-test in human infections. Transactions of the Royal Society of Tropical Medicine and Hygiene. 1991;**85**:735-738

[52] Moraes MAP, Silveira FT. Histopatologia da forma localizada de leishmaniose cutânea por *Leishmania (Leishmania) amazonensis*. Revista do Instituto de Medicina Tropical de São Paulo. 1994;**36**:459-463

[53] Silveira FT, Blackwell JM, Ishikawa EA, et al. T cell responses to crude and defined leishmanial antigens in patients from the lower Amazon region of Brazil infected with different species of *Leishmania* of the subgenera *Leishmania* and *Viannia*. Parasite Immunology. 1998;**20**:19-26

[54] Silveira FT, Lainson R, Shaw JJ, Ribeiro R. Cutaneous leishmaniasis in

Amazonia. Report of the 1st human case of mixed infection, determined by 2 different *Leishmania* species: *Leishmania braziliensis and Leishmania mexicana amazonensis*. Revista do Instituto de Medicina Tropical de São Paulo. 1984;**26**(5):272-275. DOI: 10.1590/s0036-46651984000500008

[55] Lainson R, Shaw JJ, Silveira FT, Braga RR, Ryan L, Povoa MM, et al. A *Leishmania* e as leishmanioses. In: 50 Anos de Contribuição às Ciências Biológicas e à Medicina Tropical. Belém, Pará, Brasil: Instituto Evandro Chagas; Fundação Serviços de Saúde Pública; 1986. pp. 83-124

[56] Soong L. Modulation of dendritic cell function by *Leishmania* parasites. Journal of Immunology. 2008;**180**:4355-4360

[57] Brewig N, Kissenpfennig A, Malissen B, et al. Priming of CD8+ and CD4+ T-cells in experimental leishmaniasis is initiated by different dendritic cell subtypes. Journal of Immunology. 2009;**182**:774-783

[58] Kemp K, Theander TG, Hviid L, Garfar A, Kharazmi A, Kemp M. Interferon-gamma- and tumour necrosis factor-alpha-producing cells in humans who are immune to cutaneous Leishmaniasis. Scandinavian Journal of Immunology. 1999;**49**(6):655-659. DOI: 10.1046/j.1365-3083.1999.00554.x

[59] Magalhães AV, Moraes MAP, Raick NA, Llanos-Cuentas EA, Costa JML, Cuba CC, et al. Histopatologia da leishmaniose tegumentar por *Leishmania braziliensis braziliensis*. Revista do Instituto de Medicina Tropical de São Paulo. 1986;**28**:253-262

[60] Magalhães AV, Moraes MAP, Raick NA, et al. Histopatologia da Leishmaniose Tegumentar por *Leishmania braziliensis braziliensis*: 3. Reação celular nos tecidos. Rev Inst Med Trop São Paulo. 1986;**28**:300-311

[61] Castellano LR, Filho DC, Argiro L, Dessein H, Prata A, Dessein A, et al. Th1/Th2 immune responses are associated with active cutaneous Leishmaniasis and clinical cure is associated with strong interferon-gamma production. Human Immunology. 2009;**70**(6):383-390. DOI: 10.1016/j.humimm.2009.01.007

[62] Oliveira WN, Ribeiro LE, Schrieffer A, Machado P, Carvalho EM, Bacellar O. The role of inflammatory and anti-inflammatory cytokines in the pathogenesis of human Tegumentary Leishmaniasis. Cytokine. 2014;**66**(2):127-132. DOI: 10.1016/j.cyto.2013.12.016

[63] Maspi N, Abdoli A, Ghaffarifar F. Pro- and anti-inflammatory cytokines in cutaneous Leishmaniasis: A review. Pathogens and Global. Health. 2016;**110**(6):247-260. DOI: 10.1080/20477724.2016.1232042

[64] Pacheco CMS, Araujo-Flores GV, Gonzalez K, Gomes CMC, Passero LFD, Tomokane TY, et al. Macrophage polarization in the skin lesion caused by neotropical species of *Leishmania* sp. Journal of Immunology Research. 2021;**5596876**, 8 pages. DOI: 10.1155/2021/5596876

[65] de Melo LV, Furtado RR, Campos MB, Lima LV, Vasconcelos do Santos T, Ramos PK, Stocco AC, Silveira FT. *Leishmania (V.) Lainsoni* Axenic Amastigote Antigen Was more Reactive than that of *L(V.) Braziliensis Promastigote within the Montengro Skin Test*. Cartagena, Colombia: World-Leish; 2022. pp. 1132-1133

[66] Costa JM, Marsden PD, Llanos-Cuentas EA, Netto EM, Carvalho EM, Barral A, et al.

Disseminated cutaneous Leishmaniasis in a field Clinic in Bahia, Brazil: A report of eight cases. The Journal of Tropical Medicine and Hygiene. 1986;**89**(6):319-323

[67] Galvão CE, Silva AC, Saldanha AC, Silva CM, Costa M, Costa JM. Disseminated cutaneous Leishmaniasis due to *Leishmania (V.) braziliensis* in the state of Maranhão, Brazil. Rev. Soc. Bras. Med. Trop. 1993;**26**(2):121-123. DOI: 10.1590/s0037-86821993000200008

[68] Carvalho EM, Barral A, Costa JM, Bittencourt A, Marsden P. Clinical and Immunopathological aspects of disseminated cutaneous Leishmaniasis. Acta Tropica. 1994;**56**(4):315-325. DOI: 10.1016/0001-706x(94)90103-1

[69] Turetz ML, Machado PR, Ko AI, Alves F, Bittencourt A, Almeida RP, et al. Disseminated Leishmaniasis: A new and emerging form of Leishmaniasis observed in northeastern Brazil. The Journal of Infectious Diseases. 2002;**186**(12):1829-1834. DOI: 10.1086/345772

[70] Schriefer A, Schriefer AL, Góes-Neto A, Guimarães LH, Carvalho L, Almeida RP, et al. Multiclonal *Leishmania (V.) braziliensis* population structure and its clinical implication in a region of Endemicity for American Tegumentary Leishmaniasis. Infection and Immunity. 2004;**72**(1):508-514. DOI: 10.1128/iai.72.1.508-514.2004

[71] Dantas ML, Oliveira JM, Carvalho L, Passos ST, Queiroz A, Guimaraes LH, et al. Comparative analysis of the tissue inflammatory response in human cutaneous and disseminated Leishmaniasis. Memorias do Instituto Oswaldo Cruz. 2014;**109**(2):202-209

[72] Vernal S, De Paula NA, Gomes CM, Roselino AM. Disseminated Leishmaniasis by *Leishmania Viannia* subgenus: A series of 18 cases in southeastern Brazil. Open forum. Infectious Diseases. 2026;**3**(1):ofv184. DOI: 10.1093/ofid/ofv184

[73] MembriveNA, KazumaFJ, SilveiraTGV, Teixeira JJV, Reinhold-Castro KR, Teodoro U. Disseminated cutaneous Leishmaniasis caused by *Leishmania (V.) braziliensis* in southern Brazil. Revista do Instituto de Medicina Tropical de São Paulo. 2017;**59**:e37. DOI: 10.1590/S1678-9946201759037

[74] Moriearty PL, Bittencourt AL, Pereira C, Teixeira R, Barreto E, Guimarães NA. Borderline cutaneous leishmaniasis: Clinical, immunological and histological differences from mucocutaneous leishmaniasis. Rev Inst Med Trop São Paulo. 1978;**20**(1):15-21

[75] Machado PR, Rosa ME, Costa D, Mignac M, Silva JS, Schriefer A, et al. Reappraisal of the Immunopathogenesis of disseminated Leishmaniasis: in situ and systemic immune response. Transactions of the Royal Society of Tropical Medicine and Hygiene. 2011;**105**(8):438-444. DOI: 10.1016/j.trstmh.2011.05.002

[76] Oliveira WN, Dórea AS, Carneiro PP, Nascimento MT, Carvalho LP, Machado PRL, et al. The influence of infection by different *Leishmania (Viannia) braziliensis* isolates on the pathogenesis of disseminated Leishmaniasis. Frontiers in Cellular and Infection Microbiology. 2021;**11**:740278. DOI: 10.3389/fcimb.2021.740278

[77] Queiroz A, Sousa R, Heine C, Cardoso M, Guimarães LH, Machado PR, et al. Association between an emerging disseminated form of Leishmaniasis and *Leishmania (Viannia) braziliensis* strain polymorphisms. Journal of Clinical

Microbiology. 2012;**50**(12):4028-4034. DOI: 10.1128/JCM.02064-12

[78] Gomes CMC, Silveira FT, Laurenti MD, Costa JML, Tomokane TY, Souza MGT, et al. Biological behavior of *Leishmania (Leishmania) amazonensis* isolated from different clinical forms of American tegumentar leishmaniasis in Balb/c mice. Rev Inst Med Trop São Paulo. 2003;**45**(Suppl. 13):65

[79] Machado-Coelho GL, Caiaffa WT, Genaro O, Magalhães PA, Mayrink W. Risk factors for mucosal manifestation of American cutaneous Leishmaniasis. Transactions of the Royal Society of Tropical Medicine and Hygiene. 2005;**99**(1):55-61. DOI: 10.1016/j.trstmh.2003.08.001

[80] Jirmanus L, Glesby MJ, Guimarães LH, et al. Epidemiological and clinical changes in American tegumentary leishmaniasis in an area of *Leishmania (Viannia) braziliensis* transmission over a 20-year period. The American Journal of Tropical Medicine and Hygiene. 2012;**86**:426-433

[81] Cincurá C, de Lima CMF, Machado PRL, et al. Mucosal leishmaniasis: A retrospective study of 327 cases from an endemic area of *Leishmania (V.) braziliensis*. The American Journal of Tropical Medicine and Hygiene. 2017;**97**:761-766

[82] Canário A, Queiroz M, Cunha G, Cavalcante T, Riesz V, Sharma R, et al. Presence of parasite DNA in clinically unaffected nasal mucosa during cutaneous Leishmaniasis caused by *Leishmania (Viannia) Braziliensis*. Clinical Microbiology and Infection. 2019;**25**(4):515.e5-515.e7. DOI: 10.1016/j.cmi.2018.12.027

[83] Amato VS, Tuon FF, Siqueira AM, Nicodemo AC, Neto VA. Treatment of mucosal Leishmaniasis in Latin America: Systematic review. The American Journal of Tropical Medicine and Hygiene. 2007;**77**(2):266-274. DOI: 10.4269/ajtmh.2007.77.266

[84] Goto H, Lindoso JA. Current diagnosis and treatment of cutaneous and Mucocutaneous Leishmaniasis. Expert Review of Anti-Infective Therapy. 2010;**8**(4):419-433. DOI: 10.1586/eri.10.19

[85] Gaze ST, Dutra WO, Lessa M, et al. Mucosal leishmaniasis patients display an activated inflammatory T-cell phenotype associated with a nonbalanced monocyte population. Scandinavian Journal of Immunology. 2006;**63**:70-78

[86] Cástes M, Trujilho D, Calgano M, et al. Response Th1/Th2 in human American cutaneous leishmaniasis: Its possible relevance for the design of a vaccine. Memórias do Instituto Oswaldo Cruz. 1993;**88**:42-43

[87] Da-Cruz AM, de Oliveira MP, de Luca PM, et al. Tumor necrosis fator-alpha in human American tegumentary leishmaniasis. Memórias do Instituto Oswaldo Cruz. 1996;**91**:225-229

[88] Ribeiro-de-Jesus A, Almeida RP, Lessa H, et al. Cytokine profile and pathology in human leishmaniasis. Brazilian Journal of Medical and Biological Research. 1998;**31**:143-148

[89] Blackwell JM. Tumour necrosis factor alpha and mucocutaneous leishmaniasis. Parasitology Today. 1999;**15**:73-76

[90] Christensen SM, Dillon LA. Carvalho et al. meta-transcriptome profiling of the human-*Leishmania braziliensis* cutaneous lesion. PLoS Neglected Tropical Diseases. 2016;**10**(9):e0004992. DOI: 10.1371/journal.pntd.0004992

[91] Desbois N, Pratlong F, Quist D, Dedet JP. *Leishmania (Leishmania) martiniquensis* n. sp. (Kinetoplastida: Trypanosomatidae), description of the parasite responsible for cutaneous leishmaniasis in Martinique Island (French West Indies). Parasite. 2014;**21**:1-4

[92] Shaw J, Pratlong F, Floeter-Winter L, et al. Characterization of *Leishmania (Leishmania) waltoni* n. sp. (Kinetoplastida: Trypanosomatidae) the parasite responsible for diffuse cutaneous Leishmaniasis in the Dominican Republic. The American Journal of Tropical Medicine and Hygiene. 2015;**93**:552-558

[93] Hashiguchi Y, Gomez EL, Kato H, Martini LR, Velez LN, Uezato H. Diffuse and disseminated cutaneous Leishmaniasis: Clinical cases experienced in Ecuador and a brief review. Tropical Medicine Health. 2016;**44**:2. DOI: 10.1186/s41182-016-0002-0

[94] Barral A, Costa JML, Bittencourt AL, et al. Polar and subpolar diffuse cutaneous leishmaniasis in Brazil: Clinical and immunopathologic aspects. International Journal of Dermatology. 1995;**34**:474-479

[95] Bonfim G, Nascimento C, Costa JML, et al. Variation of cytokine patterns related to therapeutic response in diffuse cutaneous leishmaniasis. Experimental Parasitology. 1996;**84**:188-194

[96] Convit J, Pinardi ME, Rondon AJ. Diffuse cutaneous leishmaniasis: A disease due to an immunological defect of the host. Transactions of the Royal Society of Tropical Medicine and Hygiene. 1972;**66**:603-610

[97] Convit J, Ulrich M, Fernandez CT, et al. The clinical and immunological spectrum of American cutaneous leishmaniasis. Transactions of the Royal Society of Tropical Medicine and Hygiene. 1993;**87**:444-448

[98] Kane MM, Mosser DM. The role of IL-10 in promoting disease progression in Leishmaniasis. Journal of Immunology. 2001;**166**(2):1141-1147. DOI: 10.4049/jimmunol.166.2.1141

[99] Costa JM, Uthant AA, Elkhoury NA, et al. Leishmaniose cutânea difusa (LCD) no Brasil após 60 anos de sua primeira descrição. Gaz Méd Bahia. 2009;**79**(Suppl. 3):16-24

[100] Bittencourt AL, Guimarães N. Imunopatologia da leishmaniose tegumentar difusa. Medicina Cutánea Ibero-Latino-Americana. 1968;**2**:395-402

[101] Bittencourt AL, Barral A. Evaluation of the histopathological classification of American cutaneous and mucocutaneous leishmaniasis. Memórias do Instituto Oswaldo Cruz. 1991;**86**:51-56

[102] Carneiro FP, De Magalhães AV, De Jesus AACM, et al. Foxp3 expression in lesions of the different clinical forms of American tegumentary leishmaniasis. Parasite Immunology. 2009;**3**:646-651

[103] Christensen SM, Belew AT, El-Sayed NM, et al. Host parasite responses in human diffuse cutaneous leishmaniasis caused by *L. amazonensis*. PLoS Neglected Tropical Diseases. 2019;**13**(3):e0007152. DOI: 10.1371/journal.pntd.0007152

[104] Levin R, Grinstein S, Canton J. The life cycle of phagosomes: Formation, maturation, and resolution. Immunological Reviews. 2016;**273**(1):156-179. DOI: 10.1111/imr.12439

[105] Uribe-Querol E, Rosales C. Control of phagocytosis by microbial pathogens.

Frontiers in Immunology. 2017;**8**:1368. DOI: 10.3389/fimmu.2017.01368

[106] de Assis RR, Ibraim IC, Nogueira PM, Soares RP, Turco SJ. Glycoconjugates in new world species of *Leishmania*: Polymorphisms in lipophosphoglycan and glycoinositolphospholipids and interaction with hosts. Biochimica et Biophysica Acta (BBA) — General Subjects. 2012;**1820**(9):1354-1365. DOI: 10.1016/J.BBAGEN.2011.11.001

[107] Carneiro MB, Peters NC. The paradox of a Phagosomal lifestyle: How innate host cell-*Leishmania amazonensis* interactions Lead to a progressive chronic disease. Frontiers in Immunology. 2021;**12**:728848. DOI: 10.3389/fimmu.2021.72884108

[108] Brittingham A, Mosser DM. Exploitation of the complement system by *Leishmania* promastigotes. Parasitology Today. 1996;**12**:444-447

[109] Giorgione JR, Turco SJ, Epand RM. Transbilayer inhibition of proteinkinase C by the lipophosphoglycan from *Leishmania donovani*. Proceedings of the National Academy Sciences USA. 1996;**93**:11634-11639

[110] Späth GF, Garraway LA, Turco SJ, Beverley SM. The role(s) of lipophosphoglycan (LPG) in the establishment of *Leishmania major* infections in mammalian hosts. Proceedings of the National Academy of Sciences USA. 2003;**100**:9536-9541

[111] Rêgo FD, Cardoso CA, Moreira POL, Nogueira PM, Araújo MS, Borges VM, et al. *Leishmania amazonensis* from distinct clinical forms/hosts has polymorphisms in Lipophosphoglycans, displays variations in immunomodulatory properties and, susceptibility to antileishmanial drugs. Cell Biology International. 2022;**46**:1-12. DOI: 10.1002/cbin.11880

[112] Farias LHS, Rodrigues APD, Silveira FT, et al. Phosphatidylserine exposure and surface sugars in two *Leishmania (Viannia) braziliensis* strains involved in cutaneous and mucocutaneous leishmaniasis. Journal of Infectious Diseases. 2013;**207**:537-543

[113] França-Costa J, Wanderley JL, Deolindo P, et al. Exposure of phosphatidylserine on *Leishmania amazonensis* isolates is associated with diffuse cutaneous leishmaniasis and parasite infectivity. PLoS One. 2012;**7**:e36595

[114] Cortez M, Huynh C, Fernandes MC, et al. *Leishmania* promotes its own virulence by inducing expression of the host immune inhibitory ligand CD200. Cell Host & Microbe. 2011;**9**:463-471

Section 3

New Insights on Leishmaniasis in Domestic Animals

Chapter 8

Feline Leishmaniasis: What Do We Know So Far?

Allana Barros Freitas, Sandra Alves Araújo, Fernando Almeida-Souza and Tatiane Aranha da Penha-Silva

Abstract

Feline leishmaniasis has a nonspecific clinical condition, like the clinical signs presented by dogs, including anorexia, lymphadenomegaly, dermatitis, emaciation, hyperthermia and atrophy of the temporal muscle. Cats have been identified as an alternative reservoir; however, felines are increasingly sought after as companion animals, further exposing man. Diagnosis in these animals is still considered a challenge, since most cats are asymptomatic. Although infected cats are less often sick than dogs, likely due to natural feline resistance, clinical illness has been linked to immunosuppressive coinfections with feline immunodeficiency virus and feline leukemia virus. The feline immune response to *Leishmania* infection has been poorly investigated. In cats, the same drugs prescribed for dogs are used in the treatment, namely pentavalent antimonials, allopurinol and miltefosine. Felines play an important role in the epidemiology of leishmaniasis, acting mainly as alternative reservoirs and accelerating the spread of the disease. Thus, there is a need for standardization of diagnostic and treatment methods, including thorough clinical evaluation. It is important to alert to the definition of public policies and to awaken feline owners, as well as the population in general, about care and prevention.

Keywords: *Leishmania*, cats, diagnoses, treatment, immune response

1. Introduction

Leishmaniasis is still one of the most neglected diseases in the world, mainly affecting developing countries. Approximately 350 million people are at risk of contracting the disease, and about two million new cases are reported annually. Annually, approximately 0.2 to 0.4 million cases of visceral leishmaniasis and 0.7 to 1.2 million cases of cutaneous leishmaniasis occur worldwide [1].

The interaction between humans and reservoir animals is an important factor in maintaining disease transmission. Although dogs are the preferred target of sandflies, in comparison to humans, studies on the susceptibility of the disease in other hosts help to elucidate more consolidated data on this zoonosis. Other important risk factors to be highlighted are the lack of basic sanitation and the urbanization process in rural areas, leading to an adequate environment for the maintenance of the vector, making the animals in these areas increasingly susceptible to infection [2].

In domestic cats, clinical disease has been associated with immunosuppressive co-infections with feline immunodeficiency virus (FIV) and feline leukemia virus (FeLV) [3]. These species are less often sick than dogs, probably due to their natural resistance. Due to the increasing number of cases of feline leishmaniasis lately in several endemic countries, the role of the domestic cat as a host of *Leishmania* has been discussed [4, 5]. Many researchers suggest the role of this species as an additional domestic reservoir [6] and/or a secondary host [7], alternative or accidental [5]. Studies carried out in Italy and Brazil showed evidence on the ability of the cat to transmit the protozoan to the vector, characterizing its behavior as a reservoir of visceral leishmaniasis [6, 8].

The feline immune response to infection has been poorly investigated. The most accepted hypothesis is that cats have natural immune resistance to arthropods and the microorganisms they transmit and that this response is controlled by genetic factors. It is possible that cats are less susceptible to the various immunomodulatory proteins contained in arthropod saliva, and cats generate protective and/or sterilizing immune responses to arthropod-borne pathogens more competently than other animals [9].

Here, we will discuss the role of domestic cats as reservoirs of leishmaniasis, as well as the immunopathological characteristics that help in the maintenance of the disease.

2. Epidemiological aspects

The etiological agents of leishmaniasis are protozoa of the genus *Leishmania* [10] in which four subgenera *Leishmania*, *Viannia, Sauroleishmania and Mundinia* are identified [11–13]. Recognition of the subgenus is performed by identifying the developmental site of the parasite in the vectors gut and by molecular characterization. Thus, the parasite is classified as subgenus *Leishmania* when they develop in the midgut and foregut and as subgenus *Viannia* when they develop in the midgut, foregut, and hindgut of the vector [11]. The subgenus *Sauroleishmania* is related to all reptile-infecting species of *Leishmania*. The subgenus Mundinia is the most recently created, in 2018, grouping *Leishmania* species that infect humans and animals, in which there is still a need for clarification about the vectors and hosts involved in its transmission cycles.

The parasite presents two main evolutive forms during its life cycle: amastigote and promastigote. The promastigote form has an elongated shape, with an external flagellum and it is found in the digestive tract of the insect vector. The amastigote is oval, with no external flagellum, and its survival depends on cells of the mammalian mononuclear phagocytic system [14]. Its biological cycle is heteroxenic, which depends on an intermediate host (invertebrate) and a definitive host (vertebrate) [15].

The transmission of leishmaniasis in felines is still poorly understood due to the few studies, but the most accepted hypothesis by researchers is the transmission by insect vectors, popularly known as sand flies [3]. The disease has vector transmission through the bite of sandflies, which in the New World belongs to the genus *Lutzomyia* and in the Old World to the genus *Phlebotomus* [16].

Only female sandflies have the hematophagous behavior that serves for the maturation of eggs in the reproductive period and become infected by taking a blood meal in a vertebrate host that develops cutaneous lesions or active parasitemia in visceral leishmaniasis. When feeding, the vector ends up ingesting infective forms of the parasite that transform into promastigotes in the midgut and reproduce in large quantities.

After this process, they move to the salivary glands, where in the next feeding of the vector, they will be inoculated together with the saliva in another vertebrate host, continuing the cycle [16, 17].

Other forms of transmission already described in dogs, mice and humans, such as horizontal transmission, have not been proven in cats [18], but blood transfusion is considered a source of infection as it has already been proven in humans and in dogs [19].

The role of cats in the leishmaniasis cycle is still poorly defined, but it is known that cats are more resistant to manifesting the disease, and anti-leishmania treatments induce clinical but not parasitological cure. Once infected, the protozoan may be available to the vector, which can make them source of sandfly infection [20, 21].

It was reported that cats could be infected by at least seven species of the genus *Leishmania*: *L. mexicana*, *Laccophilus venezuelensis*, *L. braziliensis*, *L. amazonensis*, *L. infantum* [3], *L. major* and *L. tropica* [22]. Feline leishmaniasis is distributed in 13 countries (**Figure 1**).

The first case of natural infection in a domestic feline by *Leishmania* spp. was reported in 1912 in Algeria, in an animal residing with a dog and a child with visceral leishmaniasis. After his death, amastigote forms of *Leishmania* spp. in the bone marrow of the animal confirmed the diagnosis [23].

In Brazil, the first report of leishmaniasis in felines was in 1939, in the state of Pará. A cat with lesions on the nose and ears was diagnosed with *Leishmania* spp. confirmed with cytological examination, but it was not possible to identify the species [24].

In Europe, the disease has been reported in cats in Italy [25], Spain [26], Portugal [27] and Greece [28]. In the Middle East, reports have identified the occurrence of leishmaniasis in domestic and stray cats in Turkey [22, 29], in Egypt [30] and in stray cats in Iran [31]. In Asia, the disease has been reported in cats of southeastern Thailand, in Songkhla and Satun provinces [32], in Surathani and Phangnga [33]. In North America, there were reports in the United States [34] and Mexico [35]. In South America, cases occurred in Venezuela [36] and in Brazil [37], which is currently the country with the highest number of studies on feline leishmaniasis [38].

Figure 1.
Cases of feline leishmaniasis recorded worldwide (Source: Created using the Canva app).

The atypical manifestation of the disease was reported in Africa [39] in a feline with an ocular manifestation, unilateral uveitis, which was initially classified as exudative panuveitis. After 6 months, the disease progressed and ophthalmic enucleation was required, which allowed the diagnosis by visualization of amastigote forms of *Leishmania* spp. on histopathological examination.

In Brazil, the incidence of diagnoses using molecular and serological techniques is around 7 and 8%, identifying the species *L. braziliensis*, *L. infantum* and *L. amazonensis* as the most frequent [40–42]. Studies observed the occurrence of feline leishmaniasis in 12 states [43] of the country: Mato Grosso do Sul [40, 44], Mato Grosso [44–46], Maranhão [47], Minas Gerais [48, 49], Pará [41], Paraná [50], Paraíba [51], Pernambuco [52, 53], Santa Catarina [54], São Paulo [55–57], Rio de Janeiro [58, 59], and Rio Grande do Norte [60]. The species already identified in Brazil were *L. braziliensis*, *L. amazonensis* and *L. infantum*, the latter being the one responsible for most cases that occurs in felines [43].

3. Diagnosis

The diagnosis for the detection of leishmaniasis is based on clinical findings and the epidemiology of the disease; however, laboratory analyzes are considered important tools in the diagnosis of infection [43, 61]. Currently, several methods are used (**Figure 2**), and these are applied to both dogs and cats, such as evaluation of clinical signs and parasitological, serological, and molecular tests [20, 21, 38].

Most cats are infected with *L. infantum* [51, 62]. In these cases, the predominant clinical signs are alopecia, crusts, nodules, ulcers and eye lesions, in addition to gingivostomatitis, hepatomegaly and lymphadenopathy [21]. The occurrence of signs such as fever, weight loss, stomatitis and enteritis, although rare, are also observed [55].

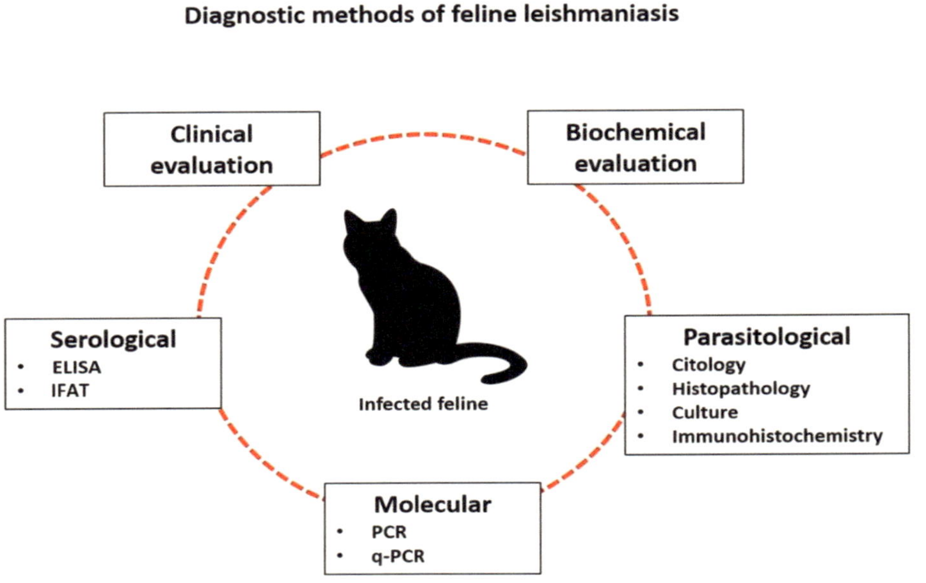

Figure 2.
Main methods of diagnosis to detect Leishmania in cats around the world.

The diagnosis of feline leishmaniasis through parasitological analysis is based on the direct observation of the amastigote forms of *Leishmania* spp. The test can be used to visualize the parasite in aspirates from lymph nodes, spleen, liver, and bone marrow [63, 64]. In animals with skin lesions, the material obtained can be used to make smears or imprints on slides, histopathological and immunohistochemical analysis and isolation in culture medium [65]. The main advantage of parasitological tests in the diagnosis of leishmaniasis is their high specificity and low cost.

Among the serological (also called immunological) methods most used in Brazil, and in the world, for the diagnosis of disease in animals are the Indirect Immunofluorescence Reaction (IFAT) and the Enzyme Immunoadsorption Assay (ELISA) [66, 67].

In Brazil, the IFAT is widely used in epidemiological studies of endemic areas. The technique uses the intact parasite as an antigen, but there is no standardized IFAT method for felines, much less a universally accepted antibody titer value that corresponds to the infection; therefore, validated titers in dogs are also used for felines [29, 68]. The ELISA is based on the reaction of antibodies present in the sera with soluble and purified *Leishmania* antigens obtained from in vitro culture [69]. In general, it is considered more sensitive and less specific than IFAT [62, 64]. Diagnosis in felines, it demonstrates greater sensitivity in clinically affected animals [70].

Although parasitological and serological methods are widely practiced, the use of molecular techniques for the detection of *Leishmania* spp. has been increasing, highlighting conventional and quantitative PCR (qPCR). These are the most used techniques for the confirmatory diagnosis of feline leishmaniasis due to the high sensitivity in the direct detection and characterization of the parasites, and especially in cases of asymptomatic animals [19, 71, 72]. PCR allows the identification of the parasite through different types of samples, such as spleen, bone marrow, lymph nodes and peripheral blood [73, 74]. In Brazil, there are reports of variations in molecular results depending on the type of tissue used for analysis [42].

The diagnosis of feline leishmaniasis is considered complex, since the number of cases of asymptomatic animals is alarming. In addition, the variety of *Leishmania* species as well as the different existing clinical manifestations have made it even more difficult, increasing the risks of human infection. In this sense, there is a great need for more specific and accurate molecular techniques.

4. Immunopathology

The immune response plays a crucial role in the control of *Leishmania* infection. T cells modulate and guide the reaction of macrophages to the parasite through the production of cytokines, although there are some differences according to the host species [75–77]. Cats are less frequently affected by arthropod-borne diseases than dogs [9].

Although these animals do not show important differences between their immune systems, the lower prevalence of *L. infantum* infection, as well as the clinical manifestations of the disease in cats, may be due to differences in the innate and adaptive immune responses of felines [3, 78]. Cats from endemic areas of canine leishmaniasis were able to activate a cell-mediated adaptative immune response against *L. infantum* through the production of IFN-γ, after stimulation with soluble parasite antigen. This response was variably associated with antibody or blood PCR positivity, providing a better estimate of cat exposure to *L. infantum* [79]. A cross-sectional comparative study of the humoral and cell-mediated adaptive immune responses to *L. infantum* of naturally exposed dogs and cats was carried out in an endemic area (southern Spain) during the sand fly season. It was observed that all dogs and cats that produced IFN-γ had low negative or positive antibodies without parasitemia, whereas animals with high parasite loads and/or antibody levels did not produce IFN-γ after stimulation with soluble *L. infantum* antigens. It was concluded that there are similarities in dogs and cats for the immunopathogenesis of infection, and that cats can mount a specific Th1 immune response against *L. infantum*. However, these responses appear to have a lower level compared to dogs [80].

A comparative study on the activation of the complement system in different hosts after stimulation with *L. infantum* confirmed lower susceptibility of cats to infection due to lower activation of the classical and alternative pathways and higher activation of the lectin pathway due to lower deposition of C4b [81].

In two necropsied cats with multicentric, nodular and ulcerated skin lesions, the presence of *Leishmania* was observed in both cytological and histopathological analyses. In addition, the presence of Mott cells and multinucleated giant cells in the dermis, lymph node and spleen were described, suggesting the excessive production of immunoglobulins which is a finding in dogs with visceral leishmaniasis. In the same work, splenic, renal and hepatic amyloidosis was found, due to immunogenic amyloid, by the production of large amounts of antibody light chains and their fragments, which can be confirmed by the presence of Mott cells [51].

L. infantum infection in cats may also be associated with immunosuppressive diseases [82]. A study described the first case of disseminated viscerocutaneous feline leishmaniasis caused by *L infantum*, associated with an invasive squamous cell carcinoma (SCC), assuming that carcinogenesis is favored by the activation and function of macrophages and dendritic cells, which may allow clones of malignant cells to escape continuously from efficient immune destruction. In addition, they observed the development of a strong humoral response to the parasite, confirmed by the high titer found in the IFAT serology [26].

5. Treatment and control

A unique treatment for feline leishmaniasis is not yet reported, so the drugs currently available for dogs are the same as those prescribed for cats. The main drugs used in the treatment of this disease are pentavalent antimonials, allopurinol and miltefosine [83–85]. Currently, the use of allopurinol is the most common treatment against feline leishmaniasis [86, 87]. In addition, the use of drug association as a treatment for leishmaniasis has been proposed for some years.

For example, the combination of meglumine antimoniate with ketoconazole was successfully administered to a feline with skin lesions [88]. In addition, the association of allopurinol and N-methyl-glucamine antimoniate in the treatment of feline leishmaniasis showed satisfactory results; however, the appearance of adverse reactions was also reported [87, 89].

In general, the treatment of leishmaniasis is considered a great challenge due to its worldwide distribution, different species of parasites, wide variety of clinical forms and immune responses [90]. Although, in most cases, the available drugs promote the full clinical recovery of the animals, they do not cause the complete elimination of the parasite; therefore, proposing preventive measures against the disease is extremely necessary.

Currently, prevention methods such as vaccines, use of a collar, spot-on and spray with repellent properties against sandflies have been used against leishmaniasis in dogs [91, 92]. However, a recent study showed that a collar initially produced for the prevention of fleas and ticks also held promise against infection in cats [93]. According to the authors, the product was 75% effective in preventing infection. Additionally, it was well tolerated by the animals, with no systemic adverse reactions and a decrease in local skin reactions. As for vaccines, some countries already use dogs; however, none are available for felines [94, 95].

6. Conclusions

The role of cats in the leishmaniasis cycle is still poorly defined, with this species being more resistant to manifesting the disease than other animals. Although there are similarities in the immunopathogenesis of the infection in dogs and cats, the response in felines seems to be at a lower level when compared to canines. However, both present treatments promote complete clinical recovery but do not cause complete elimination of the parasite. Therefore, preventive measures against the disease are still the most effective way of controlling the parasitic cycle.

Acknowledgements

The authors would like to thank FUNADESP (Agency for the Development of Private Higher Education) for the T.A. Penha-Silva scholarship. Dr. Fernando Almeida-Souza is a post-doctoral research fellow and scholarship holder of CAPES, grant number 88887.363006/2019-00.

Conflict of interest

The authors declare no conflict of interest.

Author details

Allana Barros Freitas[1], Sandra Alves Araújo[2], Fernando Almeida-Souza[1,3] and Tatiane Aranha da Penha-Silva[4*]

1 Postgraduate Program of Animal Science – UEMA, São Luís, Brazil

2 Postgraduate Program of Biotechnology, RENORBIO – UFMA, São Luís, Brazil

3 Laboratory of Protozoology, IOC-Fiocruz, Rio de Janeiro, Brazil

4 Veterinary Medicine Course, Anhanguera College, São Luís, Brazil

*Address all correspondence to: tatianearanha@hotmail.com

References

[1] WHO. Leishmaniasis. 2022. Available from: https://www.who.int/health-topics/leishmaniasis#tab=tab_1 [Accessed: February 1, 2023]

[2] Kerküklü NR, Güran M. An overview of Leishmaniasis with a focus on the island of Cyprus. Journal of Vector Borne Diseases. 2020;**57**:197-203

[3] Pennisi MG, Cardoso L, Baneth G, Bourdeau P, Koutinas A, Miró G, et al. LeishVet update and recommendations on feline leishmaniosis. Parasites & Vectors. 2015;**8**:302

[4] Duarte IRMD, Arruda CCP, Andrade ARO, Nunes VLB, Souza AI, Dourado DM, et al. Comportamento biológico de *Leishmania* (*L.*) *amazonensis* isolada de um gato doméstico (*Felis catus*) de Mato Grosso do Sul, Brasil. Revista de Patologia Tropical. 2010;**39**(1):33-40

[5] Maia C, Campino L. Can domestic cats be considered reservoir hosts of zoonotic leishmaniasis? Trends in Parasitology. 2011;**27**:341-344

[6] Silva SM, Rabelo PFB, Gontijo ND, Ribeiro RR, Melo MN, Ribeiro VM, et al. First report of infection of *Lutzomyia longipalpis* by *Leishmania* (*Leishmania*) *infantum* from a naturally infected cat of Brazil. Veterinary Parasitology. 2010;**174**(1-2):150-154

[7] Nemati T, Khanmohammadi M, Bazmani A, Mirsamadi N, Koshki MHK, Mohebali M, et al. Study on Leishmania infection in cats from Ahar, East Azerbaijan Province and Northwest Iran by parasitological, serological and molecular methods. Asian Pacific Journal of Tropical Biomedicine. 2015;**5**(1):40-43

[8] Maroli M, Pennisi MG, Di Muccio T, Khoury C, Gradoni L, Gramiccia M. Infection of Sandflies by a Cat naturally infected with *Leishmania Infantum*. Veterinary Parasitology. 2007;**145**:357-360

[9] Day MJ. Cats are not small dogs: Is there an immunological explanation for why cats are less affected by arthropod-borne disease than dogs? Parasites & Vectors. 2016;**9**:507

[10] Forattini OP. Entomologia Médica. IV: Psychodidae. Phlebotominae. Leishmanioses. Bartonelose. São Paulo, Ed. Edgard Blücher / Ed. Universidade de São Paulo, 1973:4, 658 p

[11] Lainson R, Ward RD, Shaw JJ. *Leishmania* in phlebotomid sandflies: VI. Importance of hindgut development in distinguishing between parasites of the *Leishmania mexicana* and *L. braziliensis* complexes. Proceedings of the Royal Society of London - Series B: Biological Sciences. 1977;**199**(1135):309-320

[12] Espinosa OA, Serrano MG, Camargo EP, Teixeira MMG, Shaw JJ. An appraisal of the taxonomy and nomenclature of trypanosomatids presently classified as Leishmania and Endotrypanum. Parasitology. 2018;**145**:430-442

[13] Sereno D. *Leishmania* (*Mundinia*) spp.: From description to emergence as new human and animal Leishmania pathogens. New Microbes and New Infections. 2019;**30**:100540. DOI: 10.1016/j.nmni.2019.100540

[14] Lindoso LJA, Luz GK. Leishmaniose visceral. In: Infectologia Pediátrica. São Paulo: Editora Atheneu; 2008. pp. 01-136

[15] Barbosa L, Dolabella SS, Katagiri S. Introdução à Saúde Pública. Sergipe: CESAD; 2011

[16] Neves DP. Parasitologia humana. 13ª ed. São Paulo: Atheneu; 2016

[17] Esch KJ, Petersen CA. Transmission and epidemiology of zoonotic protozoal diseases of companion animals. Clinical Microbiology Reviews. 2013;**26**(1):58-85

[18] Solano-Gallego L, Miró G, Koutinas A, Cardoso L, Pennisi MG, Ferrer L, et al. LeishVet guidelines for the practical management of canine leishmaniosis. Parasites and Vectors. 2011;**4**(1):86

[19] Pennisi MG, Hartmann K, Addie DD, Lutz H, Gruffydd-Jones T, Boucraut-Baralon C, et al. Blood transfusion in cats: ABCD guidelines for minimising risks of infectious iatrogenic complications: ABCD guidelines for minimising risks of infectious iatrogenic complications. Journal of Feline Medicine and Surgery. 2015;**17**(7):588-593

[20] Soares CSA, Duarte SC, Sousa SR. What do we know about feline leishmaniosis? Journal of Feline Medicine and Surgery. 2016;**18**(6):435-442

[21] Abramo F, Albanese F, Gattuso S, Randone A, Fileccia I, Dedola C, et al. Skin lesions in feline leishmaniosis: A systematic review. Pathogens. 2021;**10**(4):472

[22] Paşa S, Tetik Vardarlı A, Erol N, Karakuş M, Töz S, Atasoy A, et al. Detection of *Leishmania major* and *Leishmania tropica* in domestic cats in the Ege Region of Turkey. Veterinary Parasitology. 2015;**212**(3-4):389-392. DOI: 10.1016/j.vetpar.2015.07.042

[23] Sergent E, Lombard J, Quilichini M. La leishmaniose à Alger Infection simultanée d'un enfant, d'un chien et d'un chat dans la même habitation. Bulletin de la Societe de Pathologie Exotique. 1912;**5**:93-98

[24] Mello GB. Verificação da infecção natural do gato (*Felix Domesticus*) por um protozoario do genero Leishmania. Brasil Medico. 1940;**54**(12):180

[25] Spada E, Perego R, Vitale F, Bruno F, Castelli G, Tarantola G, et al. *Leishmania* spp. Infection in a non-endemic area of Northern Italy. Animals (Basel). 2020;**10**(5):817. DOI: 10.3390/ani10050817

[26] Miró G, Rupérez C, Checa R, Gálvez R, Hernández L, García M, et al. Current status of *L. infantum* infection in stray cats in the Madrid region (Spain): Implications for the recent outbreak of human leishmaniosis? Parasites & Vectors. 2014;**7**:112. DOI: 10.1186/1756-3305-7-112

[27] Maia C, Ramos C, Coimbra M, Cardoso L, Campino L. Prevalence of Dirofilaria immitis antigen and antibodies to Leishmania infantum in cats from southern Portugal. Parasitology International. 2015;**64**(2):154-156. DOI: 10.1016/j.parint.2014.11.006

[28] Giannakopoulos A, Cn T, Papadopoulos E, Spyrou V, Dc C, Valiakos G, et al. Molecular investigation and geographical distribution of *Leishmania* spp infection in stray and owned cats (Felis catus) in Thessaly, central Greece. Journal of Hellenic Veterinary Medicine Society. 2018;**68**(1):27-34

[29] Can H, Döşkaya M, Özdemir HG, Şahar EA, Karakavuk M, Pektaş B, et al. Seroprevalence of *Leishmania* infection and molecular detection of *Leishmania tropica* and *Leishmania infantu*m in stray cats of İzmir, Turkey. Experimental

Parasitology. 2016;**167**:109-114. DOI: 10.1016/j.exppara.2016.05.011

[30] Michael SA, Morsy TA, El-Seoud SF, Saleh MS. Leishmaniasis antibodies in stray cats in Ismailiya Governorate, Egypt. Journal of Egypt Society Parasitology. 1982;**12**(1):283-286

[31] Akhtardanesh B, Moeini E, Sharifi I, Saberi M, Sadeghi B, Ebrahimi M, et al. Leishmania infection in cats positive for immunodeficiency virus and feline leukemia virus in an endemic region of Iran. Veterinary Parasitology Region Studies Reports. 2020;**20**:100387. DOI: 10.1016/j.vprsr.2020.100387

[32] Junsiri W, Wongnarkpet S, Chimnoi W, Kengradomkij C, Kajeerum W, Pangjai D, et al. Seroprevalence of *Leishmania* infection in domestic animals in Songkhla and Satun provinces, southern Thailand. Tropical Biomedicine. 2017;**34**:352-362

[33] Nimsuphan B, Pangjai WC, Cheevasareechon TJ, Jirasutas W. Detection of anti-*Leishmania donovani* complex antibodies of dogs and cats from Southern Thailand. KKU Veterinary Journal. 2014;**24**:9-19

[34] Hopke K, Meyers A, Auckland L, Hamer S, Florin D, Diesel A, et al. *Leishmania mexicana* in a central Texas cat: clinical presentation, molecular identification, sandfly vector collection and novel management. JFMS Open Report. 2021;**7**(1):2055. DOI: 10.1177/2055116921999595

[35] Longoni SS, López-Cespedes A, Sánchez-Moreno M, Bolio-Gonzalez ME, Sauri-Arceo CH, Rodríguez-Vivas RI, et al. Detection of different *Leishmania* spp. and *Trypanosoma cruzi* antibodies in cats from the Yucatan Peninsula (Mexico) using an iron superoxide dismutase excreted as antigen. Comparative Immunology, Microbiology and Infectious Diseases. 2012;**35**(5):469-476. DOI: 10.1016/j.cimid.2012.04.003

[36] Bonfante-Garrido R, Urdaneta I, Urdaneta R, Alvarado J. Natural infection of cats with *Leishmania* in Barquisimeto, Venezuela. Transactions of the Royal Society of Tropical Medicine and Hygiene. 1991;**85**(1):53. DOI: 10.1016/0035-9203(91)90153-p

[37] Benassi JC, Benvenga GU, Ferreira HL, Pereira VF, Keid LB, Soares R, et al. Detection of *Leishmania infantum* DNA in conjunctival swabs of cats by quantitative real-time PCR. Experimental Parasitology. 2017 Jun;**177**:93-97. DOI: 10.1016/j.exppara.2017.04.004

[38] Asfaram S, Fakhar M, Teshnizi SH. Is the cat an important reservoir host for visceral leishmaniasis? A systematic review with meta-analysis. The Journal of Venomous Animals and Toxins Including Tropical Diseases. 2019;**25**:e20190012

[39] Verneuil M. Leishmaniose oculaire féline : à propos d'un cas. Journal francais d'ophtalmologie. 2013;**36**(4):e67-e72

[40] de Souza AI, Barros EM, Ishikawa E, Ilha IM, Marin GR, Nunes VL. Feline leishmaniasis due to *Leishmania* (*Leishmania*) *amazonensis* in Mato Grosso do Sul State, Brazil. Veterinary Parasitology. 2005;**128**(1-2):41-45. DOI: 10.1016/j.vetpar.2004.11.020

[41] Carneiro LA, Vasconcelos Dos Santos T, Lima LVDR, Ramos PKS, Campos MB, Silveira FT. First report on feline leishmaniasis caused by *Leishmania* (*Leishmania*) *amazonensis* in Amazonian Brazil. Veterinary Parasitology Region Studies Reports. 2020;**19**:100360. DOI: 10.1016/j.vprsr.2019.100360

[42] Costa-Val APD, Coura FM, Barbieri JM, Diniz L, Sampaio A, Reis JKPD, et al. Serological study of feline leishmaniasis and molecular detection of *Leishmania infantum* and *Leishmania braziliensis* in cats (*Felis catus*). Revista Brasileira de Parasitologia Veterinária. 2020;**29**(2):e003520. DOI: 10.1590/S1984-29612020023

[43] Nascimento LFJ, Cirilo TM, Gomes DS, Gomes ACA, Lima VFS, Scher R, et al. Epidemiological and diagnostic aspects of feline leishmaniasis with emphasis on Brazil: A narrative review. Parasitology Research. 2022;**121**(1):21-34. DOI: 10.1007/s00436-021-07372-9

[44] Sousa KC, Herrera HM, Domingos IH, Campos JB, Santos IM, Neves HH, et al. Serological detection of *Toxoplasma gondii*, *Leishmania infantum* and *Neospora caninum* in cats from an area endemic for leishmaniasis in Brazil. Revista Brasileira de Parasitologia Veterinária. 2014;**23**(4):449-455. DOI: 10.1590/S1984-29612014078

[45] Dahroug MA, Almeida AB, Sousa VR, Dutra V, Turbino NC, Nakazato L, et al. *Leishmania* (*Leishmania*) *chagasi* in captive wild felids in Brazil. Transactions of the Royal Society of Tropical Medicine and Hygiene. 2010;**104**(1):73-74. DOI: 10.1016/j.trstmh.2009.08.003

[46] Madruga G, Ribeiro AP, Ruiz T, Sousa VRF, Campos CG, Almeida ABPF, et al. Ocular manifestations of leishmaniasis in a cat: First case report from Brazil. Arq Bras Med Vet Zootec. 2018;**70**(5):1514-1520. DOI: 10.1590/1678-4162-9244

[47] Rocha AVVO, Moreno BFS, Cabral AD, Louzeiro NM, Miranda LM, Santos VMBD, et al. Diagnosis and epidemiology of *Leishmania infantum* in domestic cats in an endemic area of the Amazon region, Brazil. Veterinary Parasitology. 2019;**273**:80-85. DOI: 10.1016/j.vetpar.2019.08.007

[48] Passos VM, Lasmar EB, Gontijo CM, Fernandes O, Degrave W. Natural infection of a domestic cat (*Felis domesticus*) with *Leishmania* (*Viannia*) in the metropolitan region of Belo Horizonte, State of Minas Gerais, Brazil. Memórias do Instituto Oswaldo Cruz. 1996;**91**(1):19-20. DOI: 10.1590/s0074-02761996000100003

[49] Tolentino N, Pinheiro GRG, Ottino J, de Oliveira AR, Coelho CM, Tinoco HP, et al. Serological evidence of Leishmania infection by employing ELISA and rapid tests in captive felids and canids in Brazil. Veterinary Parasitology Region Studies Reports. 2019;**17**:100308. DOI: 10.1016/j.vprsr.2019.100308

[50] Matos AMRN, Caldart ET, Ferreira FP, Monteiro KC, Souza M, Brunieri DTSC, et al. Antibodies anti-trypanosomatides in domestic cats in Paraná: Who is at highest risk of infection? Revista Brasileira de Parasitologia Veterinária. 2018;**27**(2):232-236. DOI: 10.1590/s1984-296120180033

[51] Silva RBS, Portela RA, Arruda LFB, Ferreira JDS, Souto EPF, Araújo AL, et al. Natural infection by Leishmania infantum in domestic cats (Felis catus) in a municipality of moderate transmission in the Brazilian semi-arid region. Revista Brasileira de Parasitologia Veterinária. 2020;**29**(4):e016620. DOI: 10.1590/S1984-29612020102

[52] De Morais RC, Gonçalves Sda C, Costa PL, da Silva KG, da Silva FJ, Silva RP, et al. Detection of Leishmania infantum in animals and their ectoparasites by conventional PCR and real time PCR. Experimental & Applied Acarology. 2013 Apr;**59**(4):473-481. DOI: 10.1007/s10493-012-9611-4

[53] Berenguer LKAR, Gomes CFCA, Nascimento JO, Bernardi JCM, Lima VFS, de Oliveira JB, et al. Leishmania infantum infection in a domestic Cat: A Real threat or an occasional finding? Acta Parasitologica. 2021;**66**(2):673-676. DOI: 10.1007/s11686-020-00294-z

[54] Pedrassani D, Biolchi J, Gonçalves LR, Mendes NS, Zanatto DCS, Calchi AC, et al. Molecular detection of vector-borne agents in cats in Southern Brazil. Revista Brasileira de Parasitologia Veterinária. 2019;**28**(4):632-643. DOI: 10.1590/S1984-29612019077

[55] Oliveira TM, Pereira VF, Benvenga GU, Martin MF, Benassi JC, da Silva DT, et al. Conjunctival swab PCR to detect *Leishmania* spp. in cats. Revista Brasileira de Parasitologia Veterinária. 2015;**24**(2):220-222. DOI: 10.1590/S1984-29612015016

[56] Savani ES, de Oliveira Camargo MC, de Carvalho MR, Zampieri RA, dos Santos MG, D'Auria SR, et al. The first record in the Americas of an autochthonous case of Leishmania (Leishmania) infantum chagasi in a domestic cat (Felix catus) from Cotia County, São Paulo State, Brazil. Veterinary Parasitology. 2004;**120**(3):229-233. DOI: 10.1016/j.vetpar.2004.01.008

[57] Leonel JAF, Vioti G, Alves ML, Benassi JC, Silva DTD, Spada JCP, et al. Leishmaniasis in cat shelters: A serological, molecular and entomological study. Transboundary and Emerging Diseases. 2020. DOI: 10.1111/tbed.13544

[58] Schubach TM, Figueiredo FB, Pereira SA, Madeira MF, Santos IB, Andrade MV, et al. American cutaneous leishmaniasis in two cats from Rio de Janeiro, Brazil: First report of natural infection with Leishmania (Viannia) braziliensis. Transactions of the Royal Society of Tropical Medicine and Hygiene. 2004;**98**(3):165-167. DOI: 10.1016/s0035-9203(03)00040-3

[59] Figueiredo FB, Bonna IC, Nascimento LD, Costa T, Baptista C, Pacheco TM, et al. Serological evaluation for detection of anti-Leishmania antibodies in dogs and cats in the district of Santa Rita de Cássia, municipality of Barra Mansa, State of Rio de Janeiro. Revista da Sociedade Brasileira de Medicina Tropical. 2009;**42**(2):141-145. DOI: 10.1590/s0037-86822009000200009

[60] Bezerra JAB, Oliveira IVPM, Yamakawa AC, Nilsson MG, Tomaz KLR, Oliveira KDS, et al. Serological and molecular investigation of Leishmania spp. infection in cats from an area endemic for canine and human leishmaniasis in Northeast Brazil. Revista Brasileira de Parasitologia Veterinária. 2019;**28**(4):790-796. DOI: 10.1590/S1984-29612019082

[61] Torres-Guerrero E, Quintanilla-Cedillo MR, Ruiz-Esmenjaud J, Arenas R. Leishmaniasis: A review. F1000Research. 2017;**6**:750. DOI: 10.12688/f1000research.11120.1

[62] Santos NS, Pinho FA, Hlavac NRC, Nunes TL, Almeida NR, Solcà MS, et al. Feline Leishmaniasis caused by *Leishmania infantum*: Parasite sequencing, Seropositivity, and clinical characterization in an endemic area from Brazil. Frontier in Veterinary Science. 2021;**8**:734916. DOI: 10.3389/fvets.2021.734916

[63] Vides JP, Schwardt TF, Sobrinho LS, Marinho M, Laurenti MD, Biondo AW, et al. *Leishmania chagasi* infection in cats with dermatologic lesions from an endemic area of visceral leishmaniosis in Brazil. Veterinary Parasitology, Amsterdam. 2011;**178**(1-2):22-28

[64] Coura FM, Passos SKP, Pelegrino MOF, Leme FOP, Paz GF, Gontijo CMF, Costa-Val AP. Serological, molecular, and microscopic detection of *Leishmania* in cats (*Felis catus*) in Belo Horizonte, Minas Gerais State, Brazil. Revista Brasileira de Parasitologia Veterinária 2018, v. 27, n. 4, pp. 570-574. https://doi.org/10.1590/S1984-29612018005

[65] Sobrinho LS, Rossi CN, Vides JP, Braga ET, Gomes AA, Lima VMF, et al. Coinfection of *Leishmania chagasi* with *Toxoplasma gondii*, feline immunodeficiency virus (FIV) and feline Leukemia virus (FeLV) in cats from an endemic area of zoonotic visceral leishmaniasis. Veterinary Parasitology, Amsterdam. 2012;**187**(1-2):302-306

[66] Chatzis MK, Leontides L, Athanasiou LV, Papadopoulos E, Kasabalis D, Mylonakis M, et al. Evaluation of indirect immunofluorescence antibody test and enzyme-linked immunosorbent assay for the diagnosis of infection by Leishmania infantum in clinically normal and sick cats. Experimental Parasitology. 2014;**147**:54-59. DOI: 10.1016/j.exppara.2014.10.004

[67] Trevisan DAC, Lonardoni MVC, Demarchi IG. Diagnostic methods to cutaneous leishmaniasis detection in domestic dogs and cats. An. Bras. Dermatol., Rio de Janeiro. 2015;**90**(6):868-872

[68] Camprigher VM et al. Ocorrência de anticorpos anti-*Leishmania* spp. em felinos em área endêmica do estado de São Paulo. Arquivo Brasileiro de Medicina Veterinária e Zootecnia. 2019;**71**(02):439-446. DOI: 10.1590/1678-4162-10291

[69] Silveira Neto L, Marcondes M, Bilsland E, et al. Clinical and epidemiological aspects of feline leishmaniasis in Brazil. Semina: Ciências Agrárias, Londrina. 2015;**36**(3):1467-1480. DOI: 10.5433/1679-0359.2015v36n3p1467

[70] Persichetti MF, Solano-Gallego L, Vullo A, Masucci M, Marty P, Delaunay P, et al. Diagnostic performance of ELISA, IFAT and Western blot for the detection of anti-*Leishmania infantum* antibodies in cats using a Bayesian analysis without a gold standard. Parasites & Vectors. 2017;**10**(1):119

[71] Li J, Macdonald J. Advances in isothermal amplification: Novel strategies inspired by biological processes. Biosensors and Bioelectronics. 2015;**64**:196-211

[72] Pennisi MG, Persichetti MF. Feline leishmaniosis: Is the cat a small dog? Veterinary Parasitology. 2018;**251**:131-137. DOI: 10.1016/j.vetpar.2018.01.012

[73] Dourado ZF, Silva HD, Silveira-Lacerda EP, García-Zapata MTA. Panorama Histórico do Diagnóstico Laboratorial da Leishmaniose Visceral até o Surgimento dos Testes Imunocromatográficos (Rk39). Revista de Patologia Tropical / Journal of Tropical Pathology, Goiânia. 2008;**36**(3):205-214. DOI: 10.5216/rpt.v36i3.3172

[74] Moaddeb A, Behzad-Behbahani A. A simple and rapid DNA purification method for detection of *Leishmania* DNA in peripheral blood of patients with visceral Leishmaniasis. Shiraz E-Medical Journal, Shiraz, v.9, n 2, 2008

[75] Baneth G, Nachum-Biala Y, Zuberi A, Zipori-Barki N, Orshan L, Kleinerman G, et al. *Leishmania* infection in cats and dogs housed together in an animal shelter reveals a higher parasite load in infected dogs despite a greater seroprevalence among cats. Parasites Vectors. 2020;**13**:1-8

[76] Solano-Gallego L, Montserrat-Sangrà S, Ordeix L, Martínez-Orellana P. *Leishmania infantum*-specific production of ifn- and il-10 in stimulated blood from dogs with clinical leishmaniosis. Parasites Vectors. 2016;**9**:317

[77] Hosein S, Blake DP, Solano-Gallego L. Insights on adaptive and innate immunity in canine leishmaniosis. Parasitology. 2017;**144**:95-115

[78] Solano-Gallego L, Cardoso L, Pennisi MG, Petersen C, Bourdeau P, Oliva G, et al. Diagnostic challenges in the era of canine *Leishmania infantum* vaccines. Trends in Parasitology. 2017;**33**(9). DOI: 10.1016/j.pt.2017.06.004

[79] Priolo V, Martínez-Orellana P, Pennisi MG, Masucci M, Prandi D, Ippolito D, et al. Leishmania Infantum-specific IFN- production in stimulated blood from cats living in areas where canine leishmaniosis is endemic. Parasites Vectors. 2019;**12**:1-9

[80] Priolo V, Masucci M, Donato G, Solano-Gallego L, Martínez-Orellana P, Persichetti MF, et al. Association between feline immunodeficiency virus and Leishmania infantum infections in cats: A retrospective matched case-control study. Parasites Vectors. 2022;**15**:107

[81] Tirado TC, Bavia L, Ambrosio AR, Campos MP, de Almeida SM, Messias Reason IJ, et al. A comparative approach on the activation of the three complement system pathways in different hosts of visceral leishmaniasis after stimulation with Leishmania infantum. Developmental and Comparative Immunology. 2021;**120**:104061. DOI: 10.1016/j.dci.2021.104061

[82] Grevot A, Hugues P, Pratlong F, et al. Leishmaniosis due to Leishmania infantum in a FIV and FelV positive cat with a squamous cell carcinoma diagnosed with histological, serological and isoenzymatic methods. Parasite. 2005;**12**:271-275

[83] Pennisi MG, Persichetti MF, Migliazzo A, De Majo M, Iannelli NM, Vitale F. Feline leishmaniosis: Clinical signs and course in 14 followed up cases. In: Proceedings LXX Convegno SISVet; Presented at the LXX Convegno SISVet. Italy Palermo; 2016. pp. 166-167

[84] Bardagi M, Lloret A, Dalmau A, Esteban D, Font A, Leiva M, et al. Feline leishmaniosis: 15 cases. Veterinary Dermatology. 2016;**27**:112-113

[85] Brianti E, Celi N, Napoli E, Abbate JM, Arfuso F, Gaglio G, et al. Treatment and long-term follow-up of a cat with leishmaniosis. Parasites & Vectors. 2019;**12**:121

[86] Pennisi MG, Hartmann K, Lloret A, Addie D, Belák S, Boucraut-Baralon C, et al. Leishmaniosis in cats: ABCD guidelines on prevention and management. Journal of Feline Medicine and Surgery. 2013;**15**(7):638-642. DOI: 10.1177/1098612X13489229

[87] Basso MA, Marques C, Santos M, et al. Successful treatment of feline leishmaniosis using a combination of allopurinol and N-methylglucamine antimoniate. Journal of Feline Medicine and Surgery Open Reports. 2016;**1-7**. DOI: 10.1177/2055116916630002

[88] Hervás J, Chacón-M De Lara F, Sdnchez-lsarria MA, Pellicer S, Carrasco L, Castillo JA, et al. Two cases of feline visceral and cutaneous leishmaniosis in Spain. Journal of Feline Medicine and Surgery. 1999;**1**:101-105. DOI: 10.1016/S1098-612X(99)90066-9

[89] Pereira A, Valente J, Parreira R, Cristovão JM, Azinheira S, Campino L, et al. An unusual case of feline leishmaniosis with involvement of the mammary glands. Topics in Companion Animal Medicine. 2019;**37**:100356. DOI: 10.1016/j.tcam.2019.100356

[90] Sundar S, Singh A. Recent developments and future prospects in the treatment of visceral leishmaniasis. Therapeutic Advanced Infectious Diseases. 2016;**3**(3-4):98-109

[91] Otranto D, Dantas-Torres F. The prevention of canine leishmaniasis and its impact on public health. Trends in Parasitology. 2013;**29**(7):339-345. DOI: 10.1016/j.pt.2013.05.003

[92] Virginia DV, Owen H, Poapolathep A, Mario G. Natural substances as new potential strategies for the treatment of Leishmaniosis in dogs. American Journal of Animal and Veterinary Sciences. 2017;**12**(3):169-175. DOI: 10.3844/ajavsp.2017.169.175

[93] Brianti E, Falsone L, Napoli E, et al. Prevention of feline leishmaniosis with an imidacloprid 10%/flumethrin 4.5% polymer matrix collar. Parasites Vectors. 2017;**10**:334. DOI: 10.1186/s13071-017-2258-6

[94] Moreno J, Vouldoukis I, Martin V, McGahie D, Cuisinier AM, Gueguen S. Use of a LiESP/QA-21 vaccine (CaniLeish) stimulates an appropriate Th1-dominated cell-mediated immune response in dogs. PLoS Neglected Tropical Diseases. 2012;**6**:e1683

[95] Fernández Cotrina J, Iniesta V, Monroy I, Baz V, Hugnet C, Marañon F, et al. A large-scale field randomized trial demonstrates safety and efficacy of the vaccine LetiFend®against canine leishmaniosis. Vaccine. 2018;**36**:1972-1982

Chapter 9

Current Challenges in the Association between Canine Leishmaniasis and Malignant Disorders

Helen Silva Ribeiro, Flávia de Oliveira Cardoso, Ana Lucia Abreu-Silva, Kátia da Silva Calabrese and Fernando Almeida-Souza

Abstract

Neoplasms in dogs are estimated to account for up to 83% of all canine malignancies, while canine leishmaniasis is endemic and highly incident in areas like Brazil, east Africa, and India. The worldwide incidence of both diseases may indicate the presence of these two pathologies in the same tissue sample as eventual cases. However, a relationship between leishmaniasis and malignant disorders has been postulated in human and veterinary medicine. Depending on several direct or indirect factors, leishmaniasis can affect the clinical manifestation, diagnosis, therapeutic protocols, and the outcome of various malignant disorders. In addition, the similarity of the clinical presentation of cutaneous or visceral leishmaniasis with the clinical signs observed in many tumors may lead to misdiagnosis. Clinical cases of co-localization of *Leishmania* parasites and tumor cells in dogs were more associated with transmissible venereal tumors than histiocytic tumors, non-histiocytic lymphoma, or squamous cell carcinoma. This chapter reviews the circumstances involving canine leishmaniasis and malignant disorders, and the current challenge that the association between them represents to misdiagnosis or tumor-promoting potential of leishmaniasis, providing an overview of dog cases.

Keywords: *Leishmania*, neoplasm, dog, transmissible venereal tumor, diagnosis

1. Introduction

Neoplasms in dogs are estimated to account for up to 83% of all canine malignancies [1]. Its etiology is largely unknown and believed to be multifactorial [2], genetic [3, 4], auto-immune [5], hemoparasitic [6, 7], and viral diseases [8]. The exact mechanism by which chronic inflammatory processes induce oncogenesis is still

IntechOpen

unclear, but persistent immunosuppression [9], epigenetic events [10], and relevant association with immunosenescence [11] have all been associated with neoplasm developments.

Leishmania infantum-infected macrophages parasitize tissues commonly causing granulomatous inflammatory reactions [12]. Infected dogs with poor body condition, or cachexia, mainly display skin signs including skin peeling, cutaneous lesions (nodular, ulcerative, and pustular), and exfoliative dermatitis. A wide variety of clinical signs and a number of unusual presentations make diagnosis a challenge to veterinarians [13–16].

L. infantum invades most of the dogs tissues and organs [16, 17] and it has also been sporadically reported in association with different types of tumors [7, 18, 19], including cutaneous large-cell lymphoma [20], cutaneous canine transmissible venereal tumor [21], splenic [20], hemangiosarcoma [22], soft tissue sarcoma, cutaneous small-cell lymphoma, adrenocortical adenoma [7], and perianal adenoma [23].

In transmissible venereal tumor (TVT), sarcoma, small-cell lymphoma, and adrenocortical adenoma, amastigotes were observed in the macrophage's cytoplasmic vacuoles and inside the neoplastic cells, probably in consequence of an atypical dissemination of *Leishmania* parasites [23]. These findings demonstrate the ability of *Leishmania* to parasitize neoplastic cells other than leucocytes [7, 19].

A relationship between leishmaniasis and malignant disorders has been postulated in human and veterinary medicine [22, 24]. Among the proposed theories, the chronic inflammation induced by *Leishmania* persistence, the interference with the local systemic immune system, and a direct implication of the parasite in the pathogenesis of cancer are considered most important. Etiopathogenic theories are based on the pathological observation that chronic inflammation can lead to development of dysplasia, abnormalities in cell division and activation of the p53 protein (an important tumor suppressor protein, whose loss of function leads the cell to genomic instabilities, i.e., the tumor sets in) [25]. However, definitive evidence of this association has not been proven to date. Depending on several direct or indirect factors, leishmaniasis can affect the clinical manifestation, diagnosis, therapeutic protocols, course, and outcome of various malignant disorders. The association between leishmaniasis and malignancy can be identified in four distinct situations: leishmaniasis mimicking malignancy, leishmaniasis associated with malignancy, malignancy developing in patients with leishmaniasis blemish, and leishmaniasis developing in patients with malignancy [24]. This chapter reviews these situations involving canine leishmaniasis and malignant disorders, and the current challenge that the association between them represents to misdiagnosis or tumor-promoting potential of leishmaniasis, providing an overview of dog case reports, and when possible, draws a comparison with case reports in humans.

2. Histopathology, biochemical, and hematological parameters of canine leishmaniasis

The liver, spleen, and lymph nodes are the organs most affected by the disease [26], as well as the kidney and skin. In humans with visceral leishmaniasis, liver function and morphology are modified by the presence of an inflammatory process in the portal and intralobular region, locally concentrated or diffusely localized. In dogs experimentally infected by intravenous and intraperitoneal injection of amastigotes of *L. infantum*, Gonzalez et al. [27] verified that, in both forms of inoculation, the animals displayed alterations with multiple granulomas, with the visualization of the parasite, without predilection location. In addition to the *Kupffer* hypertrophy and hyperplasia are often seen in dogs naturally infected. Diffuse intralobular liver fibrosis was reported by Melo et al. [28], demonstrating that there are differences between infected and non-infected, with collagen deposition in symptomatic animals when compared to asymptomatic ones. When evaluated for the presence and size of granulomas, asymptomatic and symptomatic animals did not show differences between groups [29]. There is a granulomatous infection that only infiltrates *Kupffer* cells and hepatocytes. In chronic infections displaying different levels of severity, the granuloma was limited to the sinusoids and expanded to the portal region and the capsule, forming diffuse areas in resistant dogs. Hepatic granuloma in dogs affected with leishmaniasis is not organized [30].

Parasitism caused by *L. infantum* in the spleen can be verified, especially in the capsular and subcapsular regions and in the marginal zone, with mild intensity in asymptomatic animals, and intense parasitism in symptomatic ones [26, 31]. Studies demonstrated a reduction in regulatory T cells in the splenic tissue, indicating that they are capable of producing a greater amount of IL-10 when compared to non-infected dogs [32]. Santana et al. [29] found that susceptible dogs, that had a Montenegro intradermal reaction negative and splenic culture positive, showed smaller and fewer follicles, as well as changes in the splenic follicular organization, while severe form of leishmaniasis is associated with disorganization of white pulp [33].

In the kidney, the accumulation of immune complexes in the glomeruli can initiate glomerulonephritis and interstitial nephritis, which can lead to chronic renal failure in dogs, [34]. Santos et al. [35] evaluated histological alterations of dogs naturally infected with *Leishmania infantum* and observed amastigote form of the parasite in the kidneys and bladder of infected dogs, demonstrating that the clinical signs resulting from the urinary system alterations should be taken into account when diagnosing this disease. In fragments of the urinary bladder main alterations observed through the histopathological analysis were inflammatory reactions in the adventitial layer, characterized by moderate mononuclear infiltrate and focal perivascular infiltrate. The kidney showed glomerular degeneration and atrophy, granulomatous inflammation, apoptosis, and Bowman capsule lesions.

In general, skin lesions in dogs are focal ulcers, with raised edges, like human lesions. Other dermatological signs are alopecia, papular, and nodular dermatitis [36]. Boechat et al. [37] verified the presence of *L. (L.) infantum* amastigotes in the endometrium, ovaries, vaginal mucosa, skin, and mucosa of the vulva and mammary glands, and it was even possible to observe a greater parasite load and intense reaction inflammation in the vulva and breasts of naturally infected dogs. The number of foci and cellularity of the inflammatory infiltrates in symptomatic dogs is higher than in other groups conform seen by Verçosa et al. [38], as well as average area, perimeter, and extreme diameters of the inflammatory infiltrates were higher in symptomatic animals.

Canine leishmaniasis is also associated with alteration in hematological and biochemical parameters. Erythrogram of dogs seropositive for leishmaniasis shows moderate anemia, usually presented in the normocytic and normochromic form with non-regenerative character [39, 40]. This characteristic can be a consequent of the bone marrow invasion by the parasite, which induces inflammation and contributes to the reduction of erythrocyte production [41]. A thrombocytopenia is common and occurs due to vasculitis caused by immune complexes circulating blood, thrombocytopoiesis disorders, and platelet destruction [40].

Leukocyte response changes according to the stage of the disease [42]. Lymphopenia occurs due to the sequestration of cells in the spleen and lymph nodes, as well as by the destruction caused by *Leishmania* [43]. Other common changes such as eosinophilia and leukopenia are associated with clinical disease [44]. Lymphocytosis was observed in asymptomatic dogs [16, 45], biochemical parameters levels in serum may be higher than in dogs seronegative [43], and, in some cases, severe proteinuria may determine changes in normal serum protein values [46].

An unusual case of persistent pancytopenia caused by leishmaniasis in a human patient receiving immunomodulatory therapy for myeloma was recently described [47] and draws attention to possible similar cases in dogs. Pancytopenia in multiple myeloma is commonly caused by plasma cell infiltration and cytotoxic medications, however, fungal, bacterial, or viral infection, and treatment-associated myelodysplasia also warrant consideration. In regions where leishmaniasis occurs naturally, it should be also investigated, and in this case, the diagnosis of *Leishmania* infection and a proper therapeutic can reverse pancytopenia.

An usual alteration associated with leishmaniasis is hyperproteinemia, caused by the activation of B lymphocytes and high production of antibodies [40]. In fact, there is a decrease in the serum albumin level, in contrast to an expressive increase in the globulin levels, causing a significant increase in the levels of total plasmatic proteins during canine leishmaniasis [43, 48]. Liver function may also be normal or with changes suggestive of liver diseases [49]. It can be detected by increased activity of liver enzymes such as alkaline phosphatase (AP) and alanine aminotransferase (ALT) [49–51].

Alterations in biochemical parameters can also be associated with leishmaniasis treatment. A case report of a tumor lysis-like syndrome during therapy of visceral leishmaniasis consisting of hyperkalemia, hyperphosphatemia, hyperuricemia, and acute renal insufficiency, soon after the initiation of liposomal amphotericin B chemotherapy for severe visceral leishmaniasis, was observed in a human patient. Allopurinol therapy and intravenous fluid administration lead to a full recovery of the metabolic parameters. Awareness of this condition by veterinarians can lead to early recognition and preventive management of patients undergoing leishmaniasis treatment [52].

3. Tumor and *Leishmania* infection association

The worldwide incidence of cancer and the endemicity of leishmaniasis in areas like Brazil, east Africa, and India may indicate the presence of these two pathologies in the same tissue sample as eventual cases. Both diseases display an immunological bias in the pathology, course, and outcome of the disease progression associated with an impaired host Th1 response. Therefore, the Th2 cytokine microenvironment occurring in compartmentalized organs in progressive leishmaniasis [53] may potentially promote tumor cell proliferation as well as the other way around. In addition, the similarity of the clinical presentation of cutaneous, mucosal, or visceral leishmaniasis with lesions observed in many malignant disorders may lead to misdiagnosis [54]. Clinical cases of co-localization of *Leishmania* parasites and tumor cells in dogs were mainly associated with TVT or less frequently with histiocytic tumor, non-histiocytic lymphoma, or squamous cell carcinoma (**Figure 1**). First, we will discuss about the less frequent association of tumors with leishmaniasis, as well as misdiagnosis related to both diseases, and then the association of tumors with TVT will be reviewed.

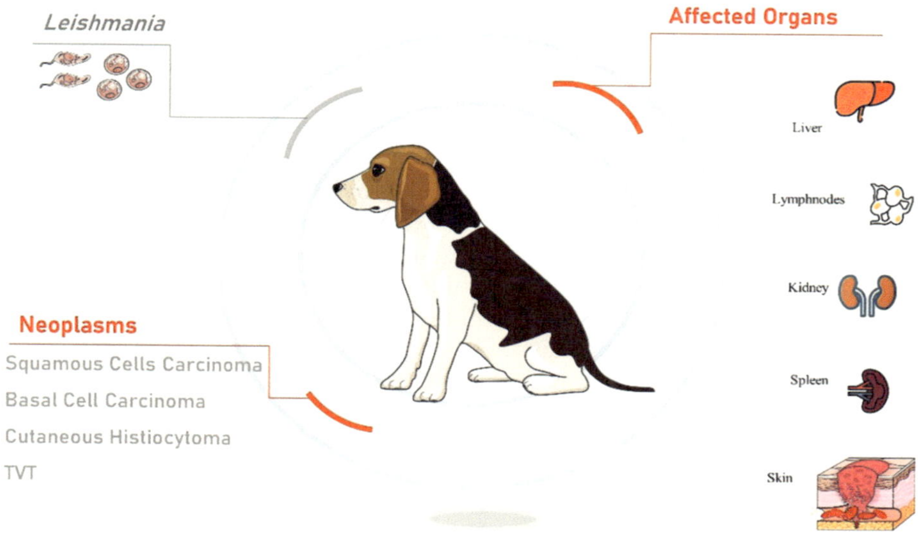

Figure 1.
Organs affected by canine leishmaniasis, and most frequent neoplasms associated with leishmaniasis in dogs.

The clinical identification of leishmaniasis is frequently related to skin lesions. However, most of the evidence about the association between cutaneous leishmaniasis and skin cancer is still based on case reports and other clinical observations rather than strong epidemiological observational studies [55]. An unlikely association of squamous cell carcinoma with an active cutaneous leishmaniasis was reported in an immunocompetent human patient [25], and a report of a case of basal cell carcinoma (Clark level IV) that had been developed on characteristic active lesion of mucocutaneous leishmaniasis, suggesting that carcinogenesis, in this case, was triggered by *Leishmania* parasites and induced by sun's ultraviolet (UV) radiation exposure [56]. Although causality is still to be proved, a biopsy may be essential to rule out associated skin malignancy in reluctant cutaneous leishmaniasis [25]. A report of a solitary cutaneous leiomyoma mimicking leishmaniasis demonstrates that neoplasms mimicking cutaneous leishmaniasis are an alert for an appropriate differential diagnosis [57].

Differential diagnosis of leishmaniasis and malignancy disorders due to similarities in clinical presentation, as well as its association, is perhaps reported for the three clinical forms of leishmaniasis: visceral, cutaneous, and mucosal. A chronic skin lesion histopathologically compatible with squamous cell carcinoma, and hence programmed for Mohs surgery, showed to be cutaneous leishmaniasis after review of the clinical and epidemiological history, and laboratory diagnosis. Treatment with miltefosine led to a complete resolution of the lesion [58]. Zambarbieri et al. [19] reported a case of *Leishmania* spp. infection in a cutaneous histiocytoma in a nodular lesion on the metacarpus of an old dog. In cytopathology examination, in some neoplastic cells, with moderate anisocytosis and anisokaryosis, *Leishmania* spp. amastigotes were observed in the cytoplasm.

A report of visceral leishmaniasis in a human patient with a lung tumor highlights leishmaniasis as a possibility to be considered in the diagnosis of causes of fever in patients undergoing chemotherapy in an endemic area [59]. In humans, visceral leishmaniasis was also reported in association with Hodgkin's lymphoma [60], melanoma

[61], and lymphoma, the last with cases of *Leishmania* infection or clinical manifestation after treatment for cancer [62]. The visceral leishmaniasis in association with multiple myeloma highlights the importance of histopathological correlation of positron emission tomography (PET) findings [63]. In this case, the liver, spleen, and bone marrow are the sites of infiltration most frequent in visceral leishmaniasis and should be observed as a differential diagnosis in PET scans.

A mucosal lesion with a tumor-like appearance by indirect laryngoscopy, but negative in indirect immunofluorescence assay test and bone marrow aspirate examination for *Leishmania*, revealed in biopsy a granulomatous inflammatory infiltrate and abundant *Leishmania* amastigotes, in addition to *Leishmania* parasites growth in culture, in a human patient. The lesions disappeared after treatment with meglumine antimoniate (850 mg/day) for 28 days [64]. Atypical cases of mucocutaneous leishmaniasis in human mimicking rectal cancer reported two different presentations: an erythematous nontender plaque measuring 5 cm by 5 cm was observed over the right perianal area of a 32-year-old male HIV patient; and two masses of different sizes and characteristics were observed around the anus and a fungating circumferential mass of 8 cm was seen above the proximal anal verge in a 40-year-old male. Atypical mucocutaneous leishmaniasis should be considered as a possible diagnosis in patients with chronic skin lesions resembling hemorrhoids and colorectal masses, especially in endemic areas for leishmaniasis [65]. A 50-year-old man mimicking multiple myeloma, presenting pancytopenia, hypergammaglobulinemia with concurrent hypoalbuminemia, showed normal cellularity, mild hemophagocytosis, increased plasma cells, and Mott cells in bone marrow aspirate. However, a thorough inspection of the bone marrow smears revealed the presence of *Leishmania* parasites. Mott cells are plasma cells with multiple cytoplasmic inclusions of immunoglobulin (Russell bodies) and it was associated with multiple myeloma cases. However, they have also been found in patients with infections such as *Leishmania* infection [66].

Other atypical report is the co-occurrence of mucosal leishmaniasis caused by *Leishmania infantum* and a retrospectively diagnosed concomitantly with a marginal zone lymphoma (MZL) in a 62-year-old man tongue. The patient was successfully treated first for MZL, but after six months, the tongue swelling returned with the identification of Leishman bodies (Giemsa stain) within the macrophages in the biopsy sample. Liposomal amphotericin B induced clinical improvement [67]. A human patient with metastatic non-small-cell lung carcinoma, after chemotherapy and a high dose of corticosteroids, showed a painful and swollen tongue. Biopsy showed an inflammation with histiocytes and *Leishmania* amastigotes indicating that the lowered immune status was important in developing mucosal leishmaniasis [68].

A gingival tumor-like lesion measuring 3 cm in the longest axis was reported to be an atypical presentation of visceral leishmaniasis in a six-year-old male mongrel dog. Upon histopathological analysis, a granulomatous inflammatory infiltrate, composed of plenty of plasma cells, lymphocytes, and histiocytes filled with amastigotes, was observed in the biopsy lesion. The diagnosis was confirmed by serological and immunohistochemistry assays. This case report reinforces that clinical and pathologist veterinarians should include leishmaniasis in the differential diagnosis of tumors and other chronic diseases of oral mucosa, especially in areas where the disease is endemic [69].

The relationship between parasites and immunotherapy is controversial, especially in leishmaniasis. It is well known that the increasing use of immunosuppressant medication by antitumor therapies makes patients susceptible to a new infection or turn symptomatic those asymptomatic *Leishmania* infected patients. On the other hand, in an experimental breast cancer model *in vivo*, *Leishmania* spp. infection demonstrated an immunostimulatory effect in cancer treatment, triggering a strong cellular immune response, determined by an increase in the production of IFN-g and IL-2 and polarized toward Th1, as necessary in the defense against tumors [70].

4. TVT and *Leishmania* infection association

Canine TVT (CTVT) is a unique neoplastic entity, contagious and sexually transmitted especially those of street life and of reproductive age, and regarded as the oldest known mammalian somatic cell neoplasm in constant transmission. CTVT occurs worldwide with higher incidence in tropical areas, and has been mostly reported in dogs (*Canis familiaris*) and foxes (*Vulpes* sp.) [71, 72]. Due to the species' communication and licking habits, it may occur in different parts of the body as around the eyes, mouth, and nasal cavity, when it is called extragenital TVT [73]. When tumors and/or their metastases occur in extragenital regions, other clinical signs may be present depending on the location of the affected organ [74–76], such as respiratory disorders, dyspnea, abdominal pain, and dysphagia [73, 77, 78].

Clinical signs often consist of genital lesions which, when present, may manifest as a single or multiple mass, friable on the foreskin or vulva [76, 77, 79], serosanguineous secretion, odor foul, and areas with or without necrosis [73]. In a case reported por Duzanski et al. [79] dog's tumor initially adhered to the adjacent tissues, and developed to a friable hyperemic mass of the cauliflower aspect. Metastasis may occur with lymphatic or visceral spread usually associated with underlying immunological impairment. In addition, TVT extragenital lesions, such as cutaneous, are common and have been reported even without primary genital lesions [71].

Atypical presentations of TVT are reported in association with *Leishmania* infection. Albanese et al. [21] reported an atypical primary cutaneous extragenital canine TVT with *Leishmania*-laden neoplastic cells in three subcutaneous round alopecic nodules, approximately 6–8 cm in diameter, located on the anterior and caudal dorsal region and in the ventral area of the neck. Intracytoplasmic *L. infantum* amastigotes were observed in the infiltrating macrophages. Intracellular amastigotes, 1–3 per cell, were present in vacuolated tumor cells. The patient received fluid therapy, erythropoietin 100 UI/kg, and intravenous injection of 0.075 mg/kg vincristine presenting regression of the nodules that decreased in size after one week, however, euthanasia was necessary due to a progressive worsening of the general condition of the dog.

A TVT was curiously reported associated with cutaneous metastasis in a female dog with leishmaniasis reactive left prescapular lymph node and subcutaneous nodular mass, not adhered to the musculature. Other visible signs were onychogryphosis, opaque, shineless, and brittle fur, with pruritic and lichenified exfoliative dermatitis in the left scapular region, ear tips and snout, and the presence of a few quantity of brownish-colored vulvar discharge with a putrid odor were the main clinical alterations. Cells with characteristics of TVT located in the genital mucosa, observed by cytology, demonstrated the presence of amastigote forms of the protozoan *Leishmania* sp. in the interior of the macrophages. TVT cells developed atypical metastasis in the cutaneous tissue in the region of the thirteenth rib. Due to the non-responsiveness of treatment and worsening of the clinical signs, the animal was euthanized [80].

Another atypical presentation of disseminated TVT associated with leishmaniasis occurred in an 8-year-old female dog that was pluriparous. There were no external genital lesions, but TVT cells were present on the serous and endometrial surface in the histopathological examination of the uterine horns, in addition to cystic endometrial hyperplasia. The history of breeding with a stray dog indicates TVT sexual transmission, and then dissemination to the omentum, peritoneum, spleen, and liver, probably favored by immunosuppression [81].

Kegler et al. [82] made the first report describing infection of tumor cells by *L. infantum* in a genital TVT from an asymptomatic 2-year-old female boxer dog, with a vaginal serosanguineous discharge. Cytological and histological examinations were performed on a friable mass occupying the upper caudal part of the vagina revealing a monomorphic population of neoplastic round cells confirming to be canine TVT, with a presence of *L. infantum* amastigotes in tumor tissue. The authors highlight the transplantation of *Leishmania*-laden neoplastic as an alternative route of venereal transmission of leishmaniasis among dogs.

Leishmania sp. amastigotes were also detected in canine TVT of a 10-year-old, mixed breed, intact female dog, with no other clinical abnormalities otherwise. Tumor tissue imprint smears, cytological, and immunohistochemistry examination revealed *Leishmania* sp. amastigotes within infiltrating macrophages, confirmed by polymerase chain reaction (PCR). Histopathology of the lesion showed extensive infiltration of lymphocytes and plasma cells within the subepithelial vaginal stroma, with numerous neutrophils and macrophages with intracytoplasmic amastigote, and rare reminiscent canine TVT neoplastic cells within reactive area of fibrosis with newly formed collagen [71]. The authors discuss about the possibility of *Leishmania* infection may have started on or from the canine venereal tumor tissue, the latter option suggesting evidence of an alternative vector-independent route of transmission for canine visceral leishmaniasis in places where both diseases coexist.

Clinicopathological findings of canine TVT in leishmaniotic dogs can aid in the understanding of the biological behavior of TVT associated with leishmaniasis for future cases. In summary, TVT can harbor a large number of *Leishmania* parasites, and canine leishmaniasis may lower the immune defense against malignancy [83].

5. Conclusions

The cases of tumors that mimic leishmaniasis lesions indicate the need for an adequate cytological and histopathological diagnosis. Likewise, cases of clinical presentation of leishmaniasis that resemble tumors require special attention and may be related to atypical *Leishmania* species in a given region, as well as host-related conditions, such as immunosuppression. The association between leishmaniasis and tumors may mean that regular diagnostic methods are not sufficiently adequate. Therefore, the use of robust diagnostic tools, with greater precision and accuracy, such as molecular tests to identify the parasite DNA through PCR, can be used to avoid a misdiagnosis. In the same way, the association between leishmaniasis and tumors is a challenge as they can arise due to cases of immunosuppression induced or not by treatment, leading to hematological and biochemical changes that can compromise the effectiveness of the treatment. Atypical conditions related to the clinical presentation of tumors or leishmaniasis in dogs require special attention, especially in endemic areas, where the probability of association of both diseases becomes greater and should always be taken into consideration in the management and conduct of the patient in human and veterinary medicine.

Acknowledgements

This research was funded by Coordenação de Aperfeiçoamento de Pessoal de Nível Superior – CAPES (Finance Code 001), by Fundação Carlos Chagas Filho de Amparo à Pesquisa do Estado do Rio de Janeiro (26/211.680/2021) and Fundação de Amparo à Pesquisa e Desenvolvimento Científico e Tecnológico do Maranhão (APP-12233/22). HSR (CNPq 150336/2023-3) and FA-S (CAPES 88887.363006/2019-00) are post-doctoral research fellows. KC (315225/2021-1) and ALA-S (313348/2021-9) are research productivity fellows by Conselho Nacional de Desenvolvimento Científico e Tecnológico - CNPq.

Author details

Helen Silva Ribeiro[1], Flávia de Oliveira Cardoso[2], Ana Lucia Abreu-Silva[1], Kátia da Silva Calabrese[2] and Fernando Almeida-Souza[1,2]*

1 Postgraduate Program of Animal Science—State University of Maranhão, São Luís, Brazil

2 Laboratory of Protozoology, Instituto Oswaldo Cruz-Fiocruz, Rio de Janeiro, Brazil

*Address all correspondence to: fernandoalsouza@gmail.com

References

[1] Vail DM, Thamm DH, Liptak JM. Hematopoietic tumors. In: Withrow and MacEwen's Small Animal Clinical Oncology. 2019. pp. 688-772. DOI: 10.1016/B978-0-323-59496-7.00033-5. Epub 2019 Nov 22. PMCID: PMC7161413

[2] Thamm DH. Novel treatments for lymphoma. Veterinary Clinics of North America: Small Animal Practice. 2019;**49**(5):903-915. DOI: 10.1016/j.cvsm.2019.04.004 Epub 2019 May 29

[3] Aricò A, Ferraresso S, Bresolin S, Marconato L, Comazzi S, Kronnie G Te, Aresu L. Array-based comparative genomic hybridization analysis reveals chromosomal copy number aberrations associated with clinical outcome in canine diffuse large B-cell lymphoma. PLoS One. 2014;**9**(11):e111817. DOI: 10.1371/journal.pone.0111817

[4] Giannuzzi D, Marconato L, Elgendy R, Ferraresso S, Scarselli E, Fariselli P, et al. Longitudinal transcriptomic and genetic landscape of radiotherapy response in canine melanoma. Veterinary and Comparative Oncology. 2019;**17**:308-316. DOI: 10.1111/vco.12473

[5] Suárez-Bonnet A, García-González B, Ramírez GA, Peñafiel-Verdú C, Jaber JR. Pemphigus erythematosus and cutaneous epitheliotropic lymphoma in a Labrador retriever dog. Iranian Journal of Veterinary Research. 2019;**20**:221-224

[6] Brunker JD, Hoover JP. B-cell lymphoma in a dog with ehrlichiosis (Ehrlichia canis) and systemic histoplasmosis (Histoplasma capsulatum). Canadian Veterinary Journal. 2007;**48**:292-295

[7] Ferro S, Palmieri C, Cavicchioli L, Zan G De, Aresu L, Benali SL. Leishmania amastigotes in neoplastic cells of 3 nonhistiocytic canine tumors. Veterinary Pathology. 2013;**50**:749-752. DOI: 10.1177/0300985813480192

[8] Bandi C, Mendoza-Roldan JA, Otranto D, Alvaro A, Louzada-Flores VN, Pajoro M, et al. Leishmania tarentolae: a vaccine platform to target dendritic cells and a surrogate pathogen for next generation vaccine research in Leishmaniases and viral infections. Parasites and Vectors. 2023;**16**:1-13. DOI: 10.1186/s13071-023-05651-1

[9] Martins LPA, Fukasawa JT, Messias MBBF, Castanho REP, Andrade AL, Sperança MA, et al. Severe visceral leishmaniasis and COVID-19 coinfection in an immunosuppressed patient. Revista do Instituto de Medicina Tropical de São Paulo. 2022;**64**:1-5

[10] Ayipo YO, Ajiboye AT, Osunniran WA, Jimoh AA, Mordi MN. Epigenetic oncogenesis, biomarkers and emerging chemotherapeutics for breast cancer. Biochimica et Biophysica Acta—Gene Regulatory Mechanisms. Oct 2022;**1865**(7):194873. DOI: 10.1016/j.bbagrm.2022.194873. Epub 2022 Sep 5. PMID: 36064110

[11] Mancuso S, Carlisi M, Santoro M, Napolitano M, Raso S, Siragusa S. Immunosenescence and lymphomagenesis. Immunity & Ageing. 2018;**15**:1-7. DOI: 10.1186/s12979-018-0130-y

[12] Morales-Yuste M, Martín-Sánchez J, Corpas-Lopez V. Canine leishmaniasis: Update on epidemiology, diagnosis, treatment, and prevention. Veterinary Science. 2022;**9**:1-20. DOI: 10.3390/vetsci9080387

[13] Corpas-López V, Merino-Espinosa G, Acedo-Sánchez C, Díaz-Sáez V,

Morillas-Márquez F, Martín-Sánchez J. Hair parasite load as a new biomarker for monitoring treatment response in canine leishmaniasis. Veterinary Parasitology. 2016;**223**:20-25. DOI: 10.1016/j.vetpar.2016.04.001

[14] Fontes JLM, Mesquita BR, Brito R, Gomes JCS, de Melo CVB, Dos Santos WLC. Anti-Leishmania infantum antibody-producing plasma cells in the spleen in canine visceral leishmaniasis. Pathogens. 2021;**10**:1-11. DOI: 10.3390/pathogens10121635

[15] Koutinas AF, Koutinas CK. Pathologic mechanisms underlying the clinical findings in canine Leishmaniosis due to leishmaniainfantum/chagasi. Veterinary Pathology. 2014;**51**:527-538. DOI: 10.1177/0300985814521248

[16] Solano-Gallego L, Koutinas A, Miró G, Cardoso L, Pennisi MG, Ferrer L, et al. Directions for the diagnosis, clinical staging, treatment and prevention of canine Leishmaniosis. Veterinary Parasitology. 2009;**165**:1-18. DOI: 10.1016/j.vetpar.2009.05.022

[17] Grano FG, José JE, Melo GD, de Souza MS, Lima VMF, Machado GF. Toll-like receptors and cytokines in the brain and in spleen of dogs with visceral Leishmaniosis. Veterinary Parasitology. 2018;**253**:30-38. DOI: 10.1016/j.vetpar.2018.02.030

[18] Di Mattia D, De Bellis F. Leishmania spp. in perianal adenoma in a dog: A case report. Topics in Companion Animal Medicine. 2019;**34**:33-35. DOI: 10.1053/j.tcam.2018.12.004

[19] Zambarbieri J, Pigoli C, Caniatti M, Scarpa P. Leishmania spp. in a cutaneous histiocytoma of an old dog. Diagnostic Cytopathology. 2021;**49**:1067-1069. DOI: 10.1002/dc.24839

[20] Manzillo VF, Pagano A, Guglielmino R, Gradoni L, Restucci B, Oliva G. Extranodal γδ-T-cell lymphoma in a dog with leishmaniasis. Veterinary Clinical Pathology. 2008;**37**:298-301. DOI: 10.1111/j.1939-165X.2008.00048.x

[21] Albanese F, Poli A, Millanta F, Abramo F. Primary cutaneous extragenital canine transmissible venereal tumour with Leishmania-laden neoplastic cells: A further suggestion of histiocytic origin? Veterinary Dermatology. 2002;**13**:243-246. DOI: 10.1046/j.1365-3164.2002.00301.x

[22] Margarito JM, Ginel PJ, Molleda JM, Moreno P, Novales M, López R. Haemangiosarcoma associated with leishmaniasis in three dogs. Veterinary Record. 1994;**134**(3):66-67. DOI: 10.1136/vr.134.3.66

[23] Di Mattia D, Fondevila D, Abramo F, Fondati A. A retrospective histopathological, immunohistochemical and molecular study of the presence of Leishmania spp. in the skin of cats with head and neck ulcerative dermatitis. Veterinary Dermatology. 2018;**29**:212-e76. DOI: 10.1111/vde.12535

[24] Kopterides P, Mourtzoukou EG, Skopelitis E, Tsavaris N, Falagas ME. Aspects of the association between leishmaniasis and malignant disorders. Transactions of the Royal Society of Tropical Medicine and Hygiene. 2007;**101**:1181-1189. DOI: 10.1016/j.trstmh.2007.08.003

[25] Ouattassi N, Titou A, Hammas N, et al. Squamous cell carcinoma associated with an active cutaneous leishmaniasis in immunocompetent patient: Case presentation of an unlikely association and literature—review. The Egyptian Journal of Otolaryngology. 2022;**38**:148. DOI: 10.1186/s43163-022-00335-6

[26] Tafuri WL, De Oliveira MR, Melo MN, Tafuri WL. Canine visceral Leishmaniosis: A remarkable histopathological picture of one case reported from Brazil. Veterinary Parasitology. 2001;**96**:203-212. DOI: 10.1016/S0304-4017(00)00436-2

[27] Gonzalez JL, Rollan E, Novoa C, Castano M. Structural and ultrastructural hepatic changes in experimental canine leishmaniasis. Histology and Histopathology. 1988;**3**:323-329

[28] Melo F, Amaral M, Oliveira P, Lima W, Andrade M, Michalick M, et al. Diffuse intralobular liver fibrosis in dogs naturally infected with Leishmania (Leishmania) chagasi. American Journal of Tropical Medicine and Hygiene. 2008;**79**:198-204. DOI: 10.4269/ajtmh.2008.79.198

[29] Santana CC, Vassallo J, De Freitas LAR, Oliveira GGS, Pontes-De-Carvalho LC, Dos-Santos WLC. Inflammation and structural changes of splenic lymphoid tissue in visceral leishmaniasis: A study on naturally infected dogs. Parasite Immunology. 2008;**30**:515-524. DOI: 10.1111/j.1365-3024.2008.01051.x

[30] Rallis T, Day MJ, Saridomichelakis MN, Adamama-Moraitou KK, Papazoglou L, Fytianou A, et al. Chronic hepatitis associated with canine Leishmaniosis (Leishmania infantum): A clinicopathological study of 26 cases. Journal of Comparative Pathology. 2005;**132**:145-152. DOI: 10.1016/j.jcpa.2004.09.004

[31] Pinto AJW, Figueiredo MM, Ferreira RA, Caliari MV, Tafuri WL. Unusual small intestine inflammatory lesions in a dog with visceral leishmaniasis. Brazilian Journal of Veterinary Pathology. 2013;**6**:19-25

[32] Silva KLO, De Andrade MMC, Melo LM, Perosso J, Vasconcelos RO, Munari DP, et al. CD4+FOXP3+ cells produce IL-10 in the spleens of dogs with visceral leishmaniasis. Veterinary Parasitology. 2014;**202**:313-318. DOI: 10.1016/j.vetpar.2014.03.010

[33] Lima IS, Silva JS, Almeida VA, Junior FGL, Souza PAN, Larangeira DF, et al. Severe clinical presentation of visceral leishmaniasis in naturally infected dogs with disruption of the splenic white pulp. PLoS One. 2014;**9**(2):e87742. DOI: 10.1371/journal.pone.0087742

[34] Manna L, Gravino AE, Picillo E, Decaro N, Buonavoglia C. Leishmania DNA quantification by real-time PCR in naturally infected dogs treated with miltefosine. Annals of the New York Academy of Sciences. 2008;**1149**:358-360. DOI: 10.1196/annals.1428.018

[35] Santos JP d, Alves LC, Ramos RAN, Pimentel D d S, Carvalho GA d, Monteiro MFM, et al. Histological changes and immunolabeling of Leishmania infantum in kidneys and urinary bladder of dogs. Revista Brasileira de Parasitologia Veterinária. 2013;**22**:420-423. DOI: 10.1590/s1984-29612013000300017

[36] Saridomichelakis MN. Advances in the pathogenesis of canine leishmaniosis: Epidemiologic and diagnostic implications. Veterinary Dermatology. Oct 2009;**20**(5-6):471-489. DOI: 10.1111/j.1365-3164.2009.00823.x. PMID: 20178485

[37] Boechat VC, Pereira SA, Júnior AAVM, Dos Santos SA, Miranda LFC, Figueiredo FB, et al. Frequency, active infection and load of Leishmania infantum and associated histological alterations in the genital tract of male and female dogs. PLoS One. 1 Sep 2020;**15**(9):e0238188. DOI: 10.1371/

journal.pone.0238188. PMID: 32870947; PMCID: PMC7462300

[38] Verçosa BL, Melo MN, Puerto HL, Mendonça IL, Vasconcelos AC. Apoptosis, inflammatory response and parasite load in skin of Leishmania (Leishmania) chagasi naturally infected dogs: A histomorphometric analysis. Veterinary Parasitology. 26 Oct 2012;**189**(2-4):162-170. DOI: 10.1016/j.vetpar.2012.04.035. Epub 2012 May 3. PMID: 22694833

[39] Almeida V, Lima I, Fraga D, Carrillo E, Moreno J, Dos-Santos WLC. Hematological changes in dogs with visceral leishmaniasis are associated with increased ifn-γ and tnf gene expression levels in the bone marrow. Microorganisms. 2021;**9**(8):1618. DOI: 10.3390/microorganisms9081618

[40] Medeiros CM d O, Melo ACC, Lima AKF, da Silva ING, de Oliveira LC, Silva MC. Perfil hematológico de cães com Leishmaniose visceral no município de Fortaleza, Ceará. Ciência Animal. 2008;**18**:43-50

[41] da Costa-Val AP, Cavalcanti RR, de Figueiredo Gontijo N, Marques Michalick MS, Alexander B, Williams P, et al. Canine visceral leishmaniasis: Relationships between clinical status, humoral immune response, haematology and Lutzomyia (Lutzomyia) longipalpis infectivity. The Veterinary Journal. 2007;**174**:636-643. DOI: 10.1016/j.tvjl.2006.11.006

[42] Gomez MA, Belew AT, Navas A, Rosales-Chilama M, Murillo J, Dillon LAL, et al. Early leukocyte responses in ex-vivo models of healing and non-healing human Leishmania (Viannia) panamensis infections. Frontiers in Cellular and Infection Microbiology. 2021;**11**:1-12. DOI: 10.3389/fcimb.2021.687607

[43] Ferreira TMV, Oliveira ATC, de Carvalho VM, Pinheiro ADN, de Carvalho Sombra TCF, Ferreira TC, et al. Leukocytes and Albumin in Canine leishmaniasis. Acta Scientiae Veterinariae. 2021;**49**:1-7. DOI: 10.22456/1679-9216.111869

[44] Baxarias M, Jornet-rius O, Donato G, Mateu C, Alcover MM, Pennisi MG, et al. Signalment, immunological and parasitological status and clinicopathological findings of Leishmania-seropositive apparently healthy dogs. Animals. 2023;**13**:1-19

[45] Nicolato RDC, De Abreu RTd, Roatt BM, Aguiar-Soares RDdO, Reis LES, et al. Clinical forms of canine visceral leishmaniasis in naturally Leishmania infantum-infected dogs and related myelogram and hemogram changes. PLoS One. 2013;**8**(12):e82947. DOI: 10.1371/journal.pone.0082947

[46] Amusategui I, Sainz A, Rodríguez F, Tesouro MA. Distribution and relationships between clinical and biopathological parameters in canine leishmaniasis. European Journal of Epidemiology. 2003;**18**:147-156. DOI: 10.1023/A:1023090929302

[47] Camilleri M, Richards H, Pomplun S, Wilson A, Checkley A, Rabin N. Leishmaniasis as an unusual cause of pancytopenia in a patient receiving immunomodulatory therapy for myeloma. British Journal of Haematology. 2020;**190**(3):305. DOI: 10.1111/bjh.16663 Epub 2020 Apr 22

[48] Giunchetti RC, Mayrink W, Carneiro CM, Corrêa-Oliveira R, Martins-Filho OA, Marques MJ, et al. Histopathological and immunohistochemical investigations of the hepatic compartment associated with parasitism and serum biochemical changes in canine visceral leishmaniasis. Research in Veterinary Science. 2008;**84**:269-277. DOI: 10.1016/j.rvsc.2007.04.020

[49] Tesfanchal B, Gebremichail G, Belay G, Gebremariam G, Teklehaimanot G, Haileslasie H, et al. Alteration of clinical chemistry parameters among visceral leishmaniasis patients in western tigrai, ethiopia, 2018/2019: A comparative cross-sectional study. Infection and Drug Resistance. 2020;**13**:3055-3062. DOI: 10.2147/IDR.S261698

[50] Abreu-Silva AL, Lima TB, Macedo AA, Moraes-Junior FJ, Dias EL, Batista ZS, et al. Flebotomíneos Em Uma Área Endêmica Na Ilha De São Luís. Revista Brasileira de Parasitologia Veterinária. 2008;**17**:197-203

[51] Kost W d O, Pereira SA, Figueiredo FB, Mendes Junior AAV, Madeira M d F, Miranda L d FC, et al. Frequency of detection and load of amastigotes in the pancreas of Leishmania infantum-seropositive dogs: clinical signs and histological changes. Parasites and Vectors. 2021;**14**:1-13. DOI: 10.1186/s13071-021-04813-3

[52] Liberopoulos E, Alexandridis G, Elisaf M. A tumor lysis-like syndrome during therapy of visceral leishmaniasis. Annals of Clinical & Laboratory Science. 2002;**32**(4):419-421

[53] Cardoso FO, Zaverucha-do-Valle T, Almeida-Souza F, Abreu-Silva AL, Calabrese KDS. Modulation of cytokines and extracellular matrix proteins expression by Leishmania Amazonensis in susceptible and resistant mice. Frontiers in Microbiology. 2020;**11**:1986. DOI: 10.3389/fmicb.2020.01986

[54] Schwing A, Pomares C, Majoor A, Boyer L, Marty P, Michel G. Leishmania infection: Misdiagnosis as cancer and tumor-promoting potential. Acta Tropica. 2019;**197**:104855. DOI: 10.1016/j.actatropica.2018.12.010 Epub 2018 Dec 7

[55] Carrillo-Larco RM, Acevedo-Rodriguez JG, Altez-Fernandez C, Ortiz-Acha K, Ugarte-Gil C. Is there an association between cutaneous leishmaniasis and skin cancer? A systematic review. Wellcome Open Research. 2019;**4**:110. DOI: 10.12688/wellcomeopenres.15367.1

[56] Al-Kamel MA. Basal cell carcinoma developed on an active lesion of mucocutaneous leishmaniasis: A case report from Yemen. Archives of Clinical Dermatology. 2018;**1**(1):4. DOI: 10.24983/scitemed.acd.2018.00072

[57] Guder S, Kelahmetoglu O. Solitary cutaneous leiomyoma mimicking leishmaniasis. Dermatologic Therapy. 2020;**33**(6):e14003. DOI: 10.1111/dth.14003 Epub 2020 Jul 27

[58] Prieto MD, Uribe-Restrepo AF, Arcos D, Vargas DA. Case report: Squamous cell carcinoma referred for mohs surgery found to be cutaneous leishmaniasis. The American Journal of Tropical Medicine and Hygiene. 2018;**99**(6):1537-1540. DOI: 10.4269/ajtmh.18-0243

[59] Claudio U, Alessandro O, Luca C, Jacopo V, Katia F. Visceral leishmaniasis in a patient with lung tumour: A case report. Tropical Doctor. 2019;**49**(2):147-149. DOI: 10.1177/004947 5519825781 Epub 2019 Feb 6

[60] Porto VBG, Carvalho LB, Buzo BF, Litvoc MN, Santos ACS, Rocci RA, et al. Visceral leishmaniasis caused by Leishmania (Leishmania) amazonensis associated with Hodgkin's lymphoma. Revista Do Instituto De Medicina Tropical De São Paulo. 2022;**64**:e51. DOI: 10.1590/S1678-9946202264051

[61] Dunzinger A, Datinger C, Loidl A, Walcherberger B, Lengauer RA, Mehraban N, et al. Visceral hepatic

leishmaniasis in a melanoma patient in FDG-PET. Current Medical Imaging. 2022;**18**(4):425-428. DOI: 10.2174/ 157340561 7666210714122602

[62] Kalmi G, Vignon-Pennamen MD, Ram-Wolff C, Battistella M, Lafaurie M, Bouaziz JD, et al. Visceral leishmaniasis in patients with lymphoma: Case reports and review of the literature. Medicine (Baltimore). 2020;**99**(45):e22787. DOI: 10.1097/MD. 000000 0000022787

[63] Boulvard-Chollet X, Cañete-Sánchez FM, Hernández-Pérez PM, Romero-Robles LG, Garrastachu-Zumarán P. Visceral leishmaniasis versus multiple myeloma relapse: Spleen as the main clue. Revista Española de Medicina Nuclear e Imagen Molecular (English Edition). 2022;**41**(Suppl. 1):S64-S65. DOI: 10.1016/j.remnie.2022.04.003 Epub 2022 May 12

[64] Cobo F, Rodríguez-Granger J, Gómez-Camarasa C, Sampedro A, Aliaga-Martínez L, Navarro JM, et al. Localized mucosal leishmaniasis caused by Leishmania infantum mimicking cancer in the rhinolaryngeal region. International Journal of Infectious Diseases. Sep 2016;**50**:54-56. DOI: 10.1016/j.ijid.2016.08.003. Epub 2016 Aug 8. PMID: 27515498

[65] Fikre H, Teklehaimanot E, Mohammed R, Mengistu M, Abebe B, van Griensven J, et al. Atypical mucocutaneous leishmaniasis presentation mimicking rectal cancer. Case Reports in Infectious Diseases. 2023;**2023**:2768626. DOI: 10.1155/2023/2768626

[66] Papakonstantinou I, Koumpis E, Fytsili E, Panteli A, Milionis H, Papoudou-Bai A, et al. Leishmaniasis mimicking multiple myeloma. International Journal of Hematology. 2022;**44**(1):44-46. DOI: 10.1111/ ijlh.13667 Epub 2021 Jul 29

[67] Beudet H, L'Ollivier C, Bouabdallah R, Campana F, Bagonchy A, Lepidi H, et al. Co-occurrence of mucosal leishmaniasis caused by Leishmania infantum and mucosal-associated lymphoid tissue lymphoma. Travel Medicine and Infectious Disease. 2023;**52**:102519. DOI: 10.1016/j. tmaid.2022.102519 Epub 2022 Dec 15

[68] Chong GM, Ong DSY, de Mendonça MM, van Hellemond JJ. Painful and swollen tongue: Mucosal leishmaniasis due to Leishmania infantum. International Journal of Infectious Disease. 2021;**113**:109-112. DOI: 10.1016/j.ijid.2021.09.071 Epub 2021 Sep 29

[69] Honse CO, Figueiredo FB, Gremião IDF, Madeira MF, Alencar NX, Miranda LHM, et al. Atypical tumor-like mass of canine visceral leishmaniasis. Source Journal of Veterinary Science. 2015;**1**(1):1-7

[70] Erdogan M, Caner A, Namlıses D, Sadıqova A, Nalbantsoy A, Oltulu F, et al. Parasites and immunotherapy: Immunostimulatory effect of Leishmania spp. in cancer treatment. European Journal of Cancer. 2019;**110**:S33. DOI: 10.1016/j.ejca.2019.01.098

[71] Carreira VS, Ferrari HF, Langohr IM, Mackenzie C, Montezzo LC, Taira E, et al. Leishmania sp. amastigotes identification in canine transmissible venereal tumor. Case Reports in Veterinary Medicine. 2014;**14**:603852. DOI: 10.1155/2014/603852

[72] Ganguly B, Das U, Das AK. Canine transmissible venereal tumour: a review. Veterinary and Comparative Oncology. 2016;**14**(1):1-12. DOI: 10.1111/vco.12060 Epub 2013 Aug 25

[73] Strakova A, Murchison EP. The changing global distribution and

prevalence of canine transmissible venereal tumour. BMC Veterinary Research. 2014;**10**:168. DOI: 10.1186/s12917-014-0168-9

[74] Bendas AJR, Moreto PLDN, Coxo AB, Holguin PG, Soares DDV. Intra-abdominal transmissible venereal tumor in a dog: A case report. Brazilian Journal of Veterinary Medicine. 2022;**26**(44):e001422. DOI: 10.29374/2527-2179.bjvm001422

[75] Nabity MB, Lees GE, Boggess MM, Yerramilli M, Obare E, Yerramilli M, et al. Symmetric dimethylarginine assay validation, stability, and evaluation as a marker for the early detection of chronic kidney disease in dogs. Journal of Veterinary Internal Medicine. 2015;**29**(4):1036-1044. DOI: 10.1111/jvim.12835

[76] Oliveira CM. Doenças do sistema genital e reprodutor. In: Jericó MM, Andrade Neto JP, Kogika MM, editors. Tratado de Medicina Interna de Cães e Gatos (1). Roca; 2019

[77] Huppes RR, Silva CG, Uscategui RAR, De Nardi AB, Souza FW, Tinucci-Costa M, et al. Tumor Venéro Transmissível (TVT): Estudo retrospectivo de 144 casos. Ars Veterinária. 2014;**30**(1):13-18. DOI: 10.15361/2175-0106.2014v30n1p13-18

[78] Ostrander EA, Davis BW, Ostrander GK. Transmissible tumors: Breaking the cancer paradigm. Trends in Genetics. 2016;**32**:1-15

[79] Duzanski ADP, Feo HB, Montoya Flórez LM, Dinau FC, Paiva BR, Brandão CVS, et al. Fibrosis in canine transmissible venereal tumor after chemotherapy with vincristine. Brazilian Journal of Veterenary Medicine. 2023;**24**:e000123. DOI: 10.29374/2527-2179.bjvm000123

[80] Diamantino LUS, Oliveira AP de, Andrade K dos S, Santos MW da C, Pereira ZS, Mendonça FL de. M, et al. Transmissible venereal tumor associated with cutaneous metastasis and leishmaniasis in a bitch. Acta Scientiae Veterinariae. 2021;**49**. DOI: 10.22456/1679-9216.108475

[81] Trevizan JT, Carreira JT, Souza NC, Carvalho IR, Gomes PB, Lima VM, et al. Disseminated transmissible venereal tumour associated with leishmaniasis in a dog. Reproduction in Domestic Animals. 2012;**47**(Suppl. 6):356-358. DOI: 10.1111/rda.12110

[82] Kegler K, Habierski A, Hahn K, Amarilla SP, Seehusen F, Baumgärtner W. Vaginal canine transmissible venereal tumour associated with intra-tumoural Leishmania spp. amastigotes in an asymptomatic female dog. Journal of Comparative Pathology. 2013;**149**(2-3):156-161. DOI: 10.1016/j.jcpa.2012.11.241 Epub 2013 Jan 21

[83] Marino G, Gaglio G, Zanghì A. Clinicopathological study of canine transmissible venereal tumour in Leishmaniotic dogs. Journal of Small Animal Practice. 2012;**53**(6):323-327. DOI: 10.1111/j.1748-5827.2012.01201.x Epub 2012 Apr 10